STRANGE PARTNERS
The Story of Symbiosis

495

Dean, Anabel.
 Strange partners : the story of symbiosis / by Anabel Dean ; illustrated by L'Enc Matte. -- Minneapolis : Lerner Publications, c1976.
 96 p. : ill. ; 24 cm.

 Includes index
 Examines speci
major kinds of s
living together
organisms—
and parasi
 ISBN 0-8

3 1192 00088 2108

x577.85 Dean.A
Dean, Anabel.
Strange partners :

J POC
1608034   780205

QH548.D4 1976

AUG 8   1983

AUG 1 1 1985

DATE DUE

DEMCO. INC. 38-2931

STRANGE PARTNERS

The Story of Symbiosis

by Anabel Dean
illustrated by L'Enc Matte

Lerner Publications Company
Minneapolis, Minnesota

LIBRARY OF CONGRESS CATALOGING IN PUBLICATION DATA

Dean, Anabel.
  Strange partnerships.

  SUMMARY: Examines specific examples of three major kinds of symbiosis, or the living together of two unlike organisms—commensalism, mutualism, and parasitism.

  Includes index.

  1. Symbiosis—Juvenile literature. [1. Symbiosis] I. Matte, L'Enc. II. Title.
  QH548.D4 1976   574.5'24   75-38479
  ISBN 0-8225-1100-2

Copyright © 1976 by Lerner Publications Company

All rights reserved. International copyright secured. No part of this book may be reproduced in any form whatsoever without permission in writing from the publisher except for the inclusion of brief quotations in an acknowledged review.

Published simultaneously in Canada by J. M. Dent & Sons (Canada) Ltd., Don Mills, Ontario

Manufactured in the United States of America

International Standard Book Number: 0-8225-1100-2
Library of Congress Catalog Card Number: 75-38479

Contents

Introduction... 7

COMMENSALISM.................................... 11

  Host Provides Food................................ 12
    grazing animals and birds........................ 12
    kori bustard and carmine bee-eater............. 13
    pompadoured hornbill and guenon monkey....... 15
    army ants and birds............................. 16
    shark and pilot fish............................. 17
  Host Provides Transportation..................... 19
    shark and remora............................... 19
    dung beetle and mite larvae.................... 21
    flies and bacteria.............................. 23
    barnacles and their hosts...................... 23
  Host Provides Shelter............................ 24
    osprey and smaller birds....................... 25
    termites and parrots........................... 26
    tuatara and sooty shearwater................... 27
    black tree ants and rufous woodpecker.......... 28
    innkeeper worm and its guests.................. 30
    hat-pin urchin and urchin fish................. 31

MUTUALISM.. 33

  Between Two Animals............................. 34
    zebra and ostrich.............................. 34
    honey guide and ratel.......................... 35
    ants and aphids................................ 38
    termites and protozoa.......................... 39
    hermit crab and sea anemone.................... 40
    giant sea anemone and clown fish............... 43
    symbiotic cleaning............................. 45

Between Two Plants..50
Between Plants and Animals..................................53
    flowering plants and pollen-carrying insects............53
    birds and plants..54
    mushroom fungi and leafcutter ants......................56
    three-toed sloth and algae..............................57

PARASITISM...61
   Parasitic Bacteria..62
    plant diseases caused by parasitic bacteria.............63
    animal diseases caused by parasitic bacteria............65
   Parasitic Plants..66
    parasitic fungi...67
    parasitic seed plants...................................71
   Parasitic Animals...74
    parasitic protozoa......................................75
    parasitic worms...78
    parasitic insects, ticks, and mites.....................83

Glossary..91
Index...93

Introduction

"Birds of a feather flock together." This familiar saying has become something of a cliché, but there is much truth in it. The saying could easily be expanded, for it applies not only to birds, but also to insects, fish, and mammals. Throughout nature, in fact, members of the same species band together to help one another, working for the good of the whole group. The reasons for this banding together may include food, shelter, protection, companionship, mating, and even survival.

Large groups of birds gather together to feed. The group offers protection to the individual, for there is safety in numbers. If one member of the group is frightened and flees, the others follow. When one is hurt, the others gather around and try to help. This same sort of cooperation exists among ants, bees, beavers, prairie dogs, seals, penguins, monkeys, zebras, elephants, whales, and most other species of animals. If the members of these various species *did not* cooperate and help one another, their numbers would dwindle rapidly and they would soon die out.

So birds of a feather, or members of the same species, *do* flock together, largely because they must. But close partnerships can also exist among members of completely different species. Sometimes the partnership is between two kinds of animals, sometimes between two kinds of plants, and sometimes between a plant and an animal. These strange partnerships between unlike organisms are examples of *symbiosis* (sim-by-OH-sis). This word was made up by early scientists from two Greek words—*sym* ("together") and *bios* ("life"). Symbiosis, then, means "living together." Any two different organisms that live together—whether for food, shelter, protection, transportation, or life itself—are involved in a symbiotic (sim-bee-AHT-ik) relationship.

There are three main kinds of symbiosis: commensalism, mutualism, and parasitism. In *commensalism*, one organism is helped by the relationship while the other organism is neither helped nor harmed. In *mutualism*, both organisms benefit by the relationship. In *parasitism*, one organism is helped by the relationship while the other one is harmed, or even destroyed, by it.

Because there is some overlapping of these three types of symbiosis, it is not always possible to say that a partnership definitely belongs to one certain class. Symbiotic relationships can change over a period of time, and sometimes we cannot be sure whether a partner is being benefited or harmed, or neither. There are, nevertheless, many clear-cut examples of each type of symbiosis. In the following pages, this book will describe some of the most interesting of these strange partnerships in nature.

Perched on this rhinoceros are some cattle egrets, heronlike birds that eat the insects stirred up by rhinos and other large grazing animals.

COMMENSALISM

Commensalism, you will recall, is the term used to describe two unlike organisms living together in an arrangement benefiting only one of them. The *host* does all the giving, and the *guest*, which is usually smaller, does all the taking. The host is not harmed, but there is no advantage to it in living with a partner. The uninvited guest is there because of the benefits received from the association with the host, which merely tolerates it.

Commensal partners share food, space, and shelter. In some commensal partnerships, only food is shared. But in others, the host also provides a ride for its guest, a dwelling place, and even protection. When a small boy tags along after his larger brother, the boy's enemies will not fight him because they are afraid of his big brother. Similarly, the enemies of a small commensal guest will not often attack it when it is with a larger host.

Host Provides Food

Commensal partnerships among animals sharing food are often rather loose, one-sided arrangements. The guest hangs around the host because food is plentiful there. The host puts up with the guest and goes about its own business, being neither harmed nor benefited by its hungry commensal partner.

grazing animals and birds

Large grazing animals such as buffalo, cattle, rhinoceros, and elephants attract many insects. Birds follow these grazers to pick off the insects feeding on their dung or stirred up by their feet. Remarkably, the birds display no fear as they hop about under the feet of their huge hosts to look for insects.

One of the best examples of this type of commensalism is the partnership between the cattle egret, a small white heronlike bird of Africa, and these grazers. When you see egrets riding on the backs of cattle, elephants, rhinoceros, and other large grazing animals, you might think that the birds were there for the ride. But this is not the case. The egrets are there for food—especially insects. They ride because the broad backs of their hosts are a handy resting place. When the birds are hungry, they simply walk beside the large grazing animals and feed on the insects stirred up by the animals' hoofs. The hosts get nothing in return, but they put up with the egrets, just the same.

A carmine bee-eater rides on the back of a kori bustard, a large cranelike bird of South Africa.

kori bustard and carmine bee-eater

A smaller bird sometimes makes use of a larger bird as a "beater" to scare insects out of the grass and bushes. A curious example of this type of commensalism is the partnership between the kori bustard and the carmine bee-eater, two birds found in South Africa.

The kori bustard is a large cranelike bird that strides across the dry, flat plains of South Africa looking for seeds, plants, insects, and small rodents to eat. If frightened, it runs very rapidly and can even take off in slow flight. The bustard is the largest flying bird alive today; some of these birds weigh more than 30 pounds and are over three and a half feet tall. Because of its large size and its ability to run fast, the kori bustard seldom flies and is gradually losing the power of flight.

About the size of a starling, the carmine bee-eater is a small red bird with a bluish-green head and a stripe of the same color on the rump. Its favorite foods, as its name suggests, are bees, wasps, and other insects. The bee-eater teams up with the kori bustard in its search for food. It looks very colorful as it rides on the back of the muted brown, black, and white bustard. As insects take flight, stirred up by the big feet of the bustard, the carmine bee-eater swoops off its perch to catch the insects. Then it returns to ride on the larger bird.

Although the bee-eater seems to enjoy the ride, it does not team up with the bustard for that purpose. Instead, the smaller bird relies upon the bustard to supply it with a meal of bees and other insects. Most insects are camouflaged with protective coloring, or they hide in the grass and bushes to escape the sharp eyes of birds. Since the carmine bee-eater is too small to scare up the hiding insects by itself, it makes use of the large kori bustard to do this for it.

If the bee-eater cannot find a kori bustard, it rides on other animals and uses them as beaters. The brightly colored bird has been seen riding on the backs of storks, zebras, and cattle. In all cases, the carmine bee-eater benefits from the relationship while its host is neither benefited nor harmed.

pompadoured hornbill and guenon monkey

Another commensal relationship in which the host provides food occurs in the Congo jungles of central Africa. This time the partnership is between a bird, the pompadoured hornbill, and a monkey.

The pompadoured hornbill is so named because it is crowned with a white pompadour and has a long horny bill. The hornbill is sometimes called the "monkeybird," for it can usually be seen following a group of monkeys —guenon monkeys, in particular. These long-tailed monkeys of Africa are more active and playful than most other varieties. As they swing along from tree to tree, shaking the branches and chattering among themselves, the insects hiding in the trees are scared up by their commotion. The pompadoured hornbill usually stays a few feet below the monkeys, snapping up the fleeing insects with its large sawlike beak. In this way, the hornbill catches many night-flying insects (moths, butterflies, beetles) that normally sleep camouflaged in the trees during the daytime.

As with the carmine bee-eater, the pompadoured hornbill associates with its larger host for the sole purpose of obtaining food. This is a loose, one-sided partnership, for the guenon monkey ignores its feathered companion and receives nothing from it.

A ground hornbill makes a meal of the insects flushed out by an army of fierce driver ants on the move.

army ants and birds

In some tropical countries, army ants are feared by all types of animal life, including human beings. Some species of army ants, known in Africa as "driver ants," march across the ground, advancing in one long column. Others advance on a broad front, which may be 48 feet wide. Most living things fear the army ants, for they march in great numbers and can overpower and kill any animal in their path. At their approach, insects and other small animals flee in heedless terror.

Ground hornbills in Africa, along with some other

insect-eating birds in Latin America, use the army ants as beaters. They follow swarms of these fierce invaders, picking off the insects that are flushed out and put to flight by the ants. The birds do not eat the army ants—only the insects they flush out. Ground hornbills and other birds have learned that insects fleeing from the army ants will provide them with an easy and abundant meal. So they leave the ants—their commensal partners—alone, neither harming nor helping them.

shark and pilot fish

Commensalism is not limited to land animals. The ocean, which abounds with life, is the scene of many commensal partnerships. The unusual relationship between sharks and pilot fish has been observed by sailors for thousands of years. Yet only recently has it been understood.

Bluish in color and marked with dark vertical stripes, pilot fish are small narrow fish that depend on sharks to provide them with food. They are called pilot fish because they normally swim alongside or in front of sharks, as if leading the poor-sighted sharks through the water. But this is not the case. Although pilot fish do, in fact, swim near sharks, scientists have learned that they are *following,* not leading. They follow the sharks closely, swimming alongside their mouths so that they can gobble up scraps of food from the sharks' meals. Whenever the sharks eat, the pilot fish get all the leftovers!

These clever little fish get more than food from their commensal host. For by staying near the shark's mouth, pilot fish are protected from their enemies. Biologists are puzzled as to why the shark does not eat its striped companion, for the pilot fish in no way helps it. The spiny-finned fish may be distasteful to the shark, and its distinctive coloring may be a warning to the shark not to consider it as food. At any rate, the shark allows the pilot fish to be its traveling companion, giving it both food and protection and receiving nothing in return.

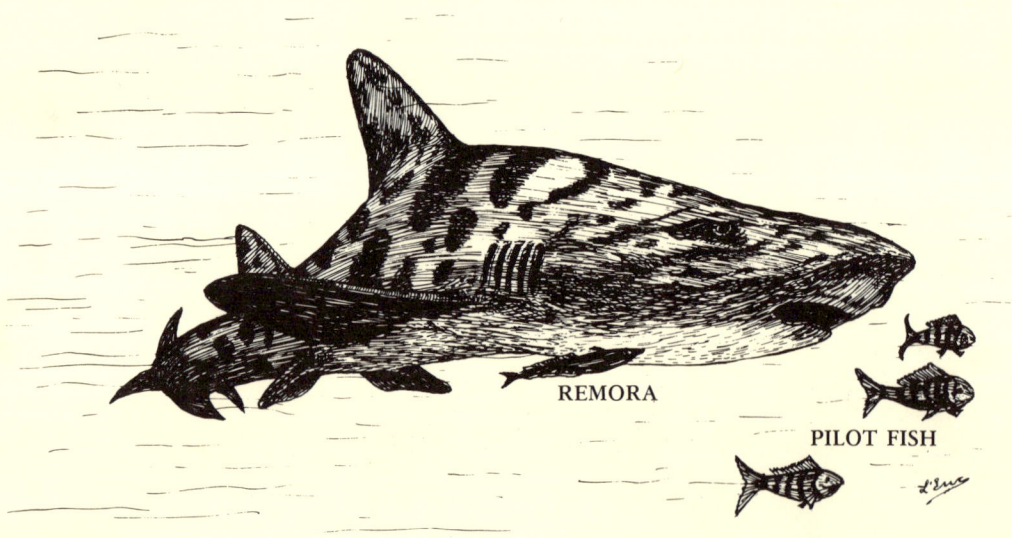

Host Provides Transportation

There are other types of commensalism besides the kind in which the host provides food. One of these is known as *phoresy* (FOR-uh-see). In this type of commensalism, one organism obtains transportation by clinging to a larger organism of a different species. The smaller of the two partners is known as the *phoront;* and the larger, the host. Sometimes the phoront hitchhikes because it is unable to travel under its own power. And sometimes the phoront is just looking for a free ride. But in all cases, the partners are not dependent on each other, and, if separated, both survive.

shark and remora

The shark has another commensal partner besides the pilot fish. This is the remora, a slender black-and-white-striped fish that grows to be two or three feet long. Whereas the pilot fish stays near the shark for the food it can obtain, the remora is there mainly for the ride. It may also snap up scraps of food dropped by the shark, however.

The remora, often called the "shark sucker," has become physically adapted to fasten itself onto the shark for a free ride. The sucker on the top of the remora's head is actually a modified dorsal fin. Resembling the sole of a tennis shoe, it is a flat, oval-shaped disc with two rows of narrow slats or suction plates running down the middle. When raised against a smooth sur-

The remora uses the oval-shaped sucker on its head to fasten itself to sharks and other large fish.

face such as the belly of a shark, the suction plates form a vacuum. It is then almost impossible to remove the remora from its host.

Since remoras are so well adapted to hitchhiking, it seems reasonable to suppose that they are poor swimmers. But this is not the case. Remoras are strong, fast swimmers who simply prefer letting sharks do the swimming for them. Sharks also provide protection, for few fish choose to come near these vicious killers.

The remora's favorite host is the shark. But the hitchhiking fish has also been known to attach itself to other large fish (especially tarpons and barracudas), to sea turtles, and even to ships. The ancient Greeks and Romans wrote of remoras attaching themselves to ships in such large numbers as to slow the ships down! The scientific name of the remora, *Echeneis naucrates*, is thus fitting, for *echeneis* means "holding back" in Latin.

Whatever they attach themselves to, remoras do no real harm. Nor do they help their hosts in any way. Luckily, the sharks and other large fish that provide them with transportation make no attempt to eat or kill the remoras.

dung beetle and mite larvae

Another example of phoresy is the strange partnership between the dung beetle and the tiny larvae (LAR-vee) of mites. The ancient Egyptians considered the dung beetle—or "scarab," as they called it—to be sacred. Likenesses of the beetle were carved or painted on buildings, and metal scarabs were used as charms. Dead scarabs have been found in large numbers in the tombs of wealthy Egyptians. The Egyptians looked upon the scarab as a symbol of immortality and believed it would help preserve the souls of the dead.

The dung beetle's name comes from its curious habit of breeding in, and feeding on, dung. The beetle, which is only about an inch long, rolls fresh dung and decayed matter into balls, and then rolls the balls to its underground breeding burrows, pushing them with its hind legs. The beetle next lays its eggs in the dung balls. Then, after the eggs turn into larvae, the hungry larvae eat the dung balls.

The industrious, hardworking scarab has many commensal partners. Among the phoronts that use the dung beetle as a means of transportation are the tiny larvae of mites. These blind, flightless creatures also live on dung. They must find a way of traveling to new supplies of dung, or they will starve to death. Somehow, they know that the dung beetle will take them to sources of food.

The mite larvae smell with special organs located on their legs, and they use these organs to sniff out

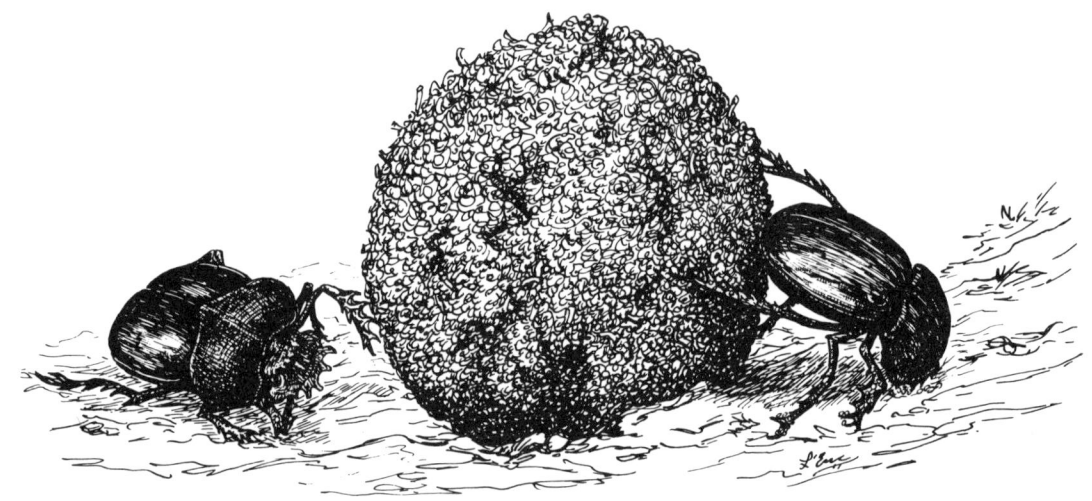
Two scarabs rolling a ball of dung

dung beetles. Once the larvae have located a dung beetle, they cling to its legs and underside as the beetle flies to a new source of dung. After the beetle lands, the larvae crawl off their host and feed on the dung. When the food supply is gone, the hungry hitchhikers seek out another ride to a new supply of dung.

Tiny worms also ride on the dung beetle to sources of food. The worms are blind like the mite larvae, but they locate the beetle by touch rather than by smell. After boarding the dung beetle, the worms secrete a liquid that hardens like glue and keeps the worms from falling off during the beetle's flight. When the beetle arrives at a fresh pile of dung, the moisture in the dung loosens the glue, allowing the worms to crawl off the beetle and enjoy the feast.

Neither the worms nor the mite larvae help the dung beetle in any way; they are just stowaways on the flight. But the beetle is not harmed by the tiny phoronts, and it doesn't seem to mind giving them a lift.

flies and bacteria

One of the most common phoretic relationships cannot even be seen with the naked eye. This is the partnership between flies and bacteria, one-celled organisms that are too small to be seen without a microscope.

Bacteria are the most numerous of all living things. Whereas some scientists consider them to be plants, others consider them to be animals. Neither view is wrong, for bacteria can have the qualities of both plants and animals. Some kinds of bacteria are harmful, causing diseases in human beings and other animals. But other kinds are helpful, such as the ones that break down dead and decaying matter into food that can be used by plants and animals.

Partly because they are so small, bacteria cannot travel long distances by themselves. But when flies walk around in garbage and manure, as they are apt to do, the bacteria found there often cling to the flies' hairy legs. In this way, the bacteria are carried many miles from their original homes. Since the flies cannot see their microscopic passengers, they are unaware of the service they are performing.

barnacles and their hosts

Barnacles are well-known saltwater phoronts that gain free transportation from a wide variety of hosts. These small shellfish can swim about freely during the first stages of their life. But later, they attach themselves permanently to underwater objects and organisms. Hard shells or boxes then grow around them, cementing the barnacles to their hosts. The barnacles are not entirely trapped in their shells, however, for a lidlike opening allows their tentacles to reach out and draw in microscopic organisms for food.

Some kinds of barnacles attach themselves only to inanimate objects such as pebbles, rocks, wharf pilings, and the bottoms of ships. Other types of barnacles prefer living hosts, largely because these hosts can provide them with transportation and so help them in their search for food. Among the favorite hosts of barnacles are whales, turtles, crabs, water beetles, oysters, and mussels. Most barnacles are not overly fussy, and will attach themselves to any of these hosts. But a few types of barnacles are quite selective, and will seek out one certain host from all the others.

Barnacles do their hosts no harm, except for slowing them down a bit with their added weight. On the other hand, barnacles do not help their hosts in any way. Theirs is strictly a one-sided relationship in which they reap all the rewards while their hosts gain nothing. Nothing, that is, but the most steadfast traveling companions in the sea!

Host Provides Shelter

A third type of commensalism, known as *synoecy* (suh-NEE-see), involves the sharing of shelter. Usually a tenant moves in with a larger host and so gets a free home. Sometimes food and protection are also furnished. The host gets nothing in return, but it makes no attempt to harm or evict its guests. Instead, it merely tolerates their presence and goes about its daily activities as if they weren't there.

osprey and smaller birds

There are not many synoecious partnerships among mammals. Several species of birds, though, have found ways to move in with other animals and so obtain free housing.

One bird, the osprey or fish hawk, is sometimes the landlord to several uninvited guests. This large bird of prey builds a huge platform of sticks on some high spot such as a treetop, a telephone pole, or the ledge of a cliff. The osprey uses the same nest for many years, building it up higher and higher each year.

This towering pile of sticks attracts smaller birds like sparrows, grackles, wrens, and black-crowned night herons. These birds make their nests in the spaces below the osprey's nest. They gain protection from living close to the osprey, for this fierce hunter is feared by most other birds of prey. The osprey's tenants have nothing to fear, however, for the osprey feeds mainly on fish, leaving smaller birds alone.

In addition to shelter and protection, the osprey's guests sometimes obtain food. When the osprey leaves its nest to go fishing, the smaller birds rummage for scraps of fish left behind in the osprey's nest.

termites and parrots

In many cases where one organism provides another with shelter, the host is larger than the guest. But this does not always hold true. For example, the paradise parrot and several other species of Australian parrots have moved in with termites! In this case, the guests are many times larger than their hosts.

Some species of termites found in Australia build huge clay mounds, often as high as 15 feet, for their nests. In turn, five or six species of Australian parrots dig tunnels into these termite mounds and hollow out places for their own nests. The parrots line their nests with special insulating material from the tunnels of the termite mounds. Then the termites bring clay and wall off the birds' nests in order to protect their own nests from any further intrusions.

Since parrots feed only on seeds and grain, they do not eat the termites. They inadvertently destroy many termite eggs, however, while digging their tunnels through the termite nests. Although the termites do not benefit from this arrangement, they make no attempt to drive out the uninvited guests. Nor do they harm the baby parrots being raised within their mounds.

The sooty shearwater shares its underground burrow with the tuatara, a fierce-looking reptile too lazy to dig its own burrow.

tuatara and sooty shearwater

New Zealand is the scene of another fascinating synoecious partnership. In this case, a large lizardlike reptile, the tuatara, sets up housekeeping in the burrow of a small gray sea bird, the sooty shearwater. These two animals are unlikely roommates, but they get along surprisingly well.

One of the rarest creatures on earth, the tuatara is the only survivor of a group of reptiles called the rhynchocephalians, or "beak-heads." Its ancestors go all the way back to the time of the dinosaurs, and the tuatara itself has existed with only minor changes for millions of years. This "living fossil" resembles an iguana, except that it has a third eye on the top of its head. About two and a half feet long, the tuatara is olive-green in color, with small yellow specks on its sides and a row of short yellow spines down the middle of its back. If it weren't for its size, this fierce-looking insect-eater might well be mistaken for a dragon!

There are some 10,000 tuataras in all, and they live on about 20 small islands off the coast of New Zealand. These cool, misty islands are also the home of many sea birds. One of these is the sooty shearwater. This small gray bird digs an underground burrow in which it lays its eggs, raises its young, and sleeps. Like the shearwater, the tuatara is accustomed to living in an underground home. But since it is too lazy and sluggish to dig its own burrow, it moves in with the sooty shearwater. This home-sharing arrangement works out quite well, for the bird and the reptile are rarely in the burrow at the same time. The sooty shearwater spends the day fishing while the tuatara, a nocturnal creature, sleeps. At night, the tuatara comes out to hunt for insects while the sooty shearwater sleeps.

The tuatara does not bother the eggs or young of its host, and vice versa. In winter, when the sooty shearwater flies north to Greenland, the tuatara remains in the burrow and hibernates.

black tree ants and rufous woodpecker

In India and Ceylon, a woodpecker and an ant have a peculiar home-sharing arrangement that has been puzzling scientists for years. Black tree ants living in these tropical countries build large football-shaped nests in the trees. Oddly enough, these ant nests are the favorite nesting places of the rufous woodpecker of southern Asia. In the spring, when the woodpecker's nesting season begins, the bird seeks out one of the ant nests. With its strong, chisel-shaped bill, the woodpecker drills into the side of the nest and hollows out a six-inch-wide hole. The woodpecker then lays its eggs in this hole and waits for them to hatch, using the ants' nest as a nesting place for its own young.

The most puzzling part of this arrangement is that the rufous woodpecker and the black tree ants are enemies! At most times of the year, the woodpecker eats the ants, and the ants drive the woodpecker away by attacking and stinging it. Once the woodpecker's nesting season arrives, however, a truce is declared. Although the rufous woodpecker accidentally destroys some ant larvae while hollowing out its hole, it does not eat any of the black tree ants so long as its eggs and babies are in the ants' nest. As for the ants, they make no attempt to stop the woodpecker from making its nest in their home. Nor do they eat the woodpecker's eggs or harm its young.

The reasons for this truce have not yet been determined. Until they are, the strange home-sharing arrangement between the rufous woodpecker and the black tree ants will remain one of nature's most baffling mysteries.

innkeeper worm and its guests

In the ocean, there are many synoecious partnerships between hosts and uninvited guests. One foot-long marine worm that lives in the mud flats along the seacoast of California has been given the name *Urechis caupo*. *Caupo* means "innkeeper" in Latin, and that is just what this worm is.

Early in life, the innkeeper worm is free swimming. But it soon digs a U-shaped burrow in the mud and settles down there for life. In order to obtain food, the worm spins a slime net over its head and fastens one end of it to one of the two burrow openings. Then the worm pumps water in through one end of the burrow and out the other. As this pumping action occurs, the worm's net strains out tiny organisms in the water. When the net is full of food, the worm swallows both the net and the food. Then the worm spins another net and starts collecting food again.

The pumping action of the innkeeper worm causes sea water containing tiny specks of food to circulate

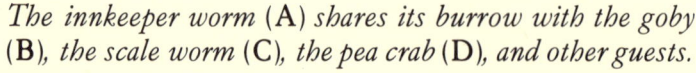

The innkeeper worm (A) shares its burrow with the goby (B), the scale worm (C), the pea crab (D), and other guests.

throughout its burrow. Many uninvited guests move into the burrow to take advantage of this. A generous host, the innkeeper worm provides them with shelter, protection, and food. Among the most common guests of the innkeeper worm are the pea crab, the scale worm, and the goby, a small spiny-finned fish.

The goby hovers just inside the burrow's entrance for shelter and protection, but it forages outside the burrow for its food. When danger threatens, all the guests of the innkeeper worm huddle close to the larger host for protection. They offer nothing in return, but the innkeeper worm does not seem to mind sharing its home with them.

hat-pin urchin and urchin fish

Another synoecious partnership exists between the hat-pin urchin of the tropics and certain small thin fish. As its name suggests, the hat-pin urchin is covered with long sharp needles or spines that stick out from its ball-shaped body in all directions. Most fish avoid the sea urchin, for its spines are movable and can cause injury or death. But two species of tropical fish team up with the urchin, unafraid of its spines. These are the shrimpfish and the clingfish. Because of their close association with the sea urchin, these small slender fish are often called "urchin fish."

The urchin fish stay near the hat-pin urchin and depend on it for shelter and protection. They forage for their own food; but at the first sign of danger, they scurry back to the urchin. Then they slide down among the urchin's spines, with their noses close to the urchin's body and their tails pointing outward. To avoid being stabbed, the fish sway back and forth with the movements of the urchin's spines. So long as the little urchin fish remain in this position, surrounded by the sharp spines of their host, large predatory fish will not try to attack them.

Here again, as with all the other examples in this chapter, the guests benefit from the relationship while their commensal host is neither benefited nor harmed by it.

MUTUALISM

Mutualism, or the living together of two unlike organisms for mutual aid, is the type of symbiosis seen most often in nature. When mutualism occurs, *both partners* benefit from the arrangement. Mutualism is a tighter form of symbiosis than commensalism, for it involves more teamwork and cooperation. Whereas commensalism is a one-sided relationship in which the host gives and the guest takes, mutualism is a two-sided relationship in which both members are full partners, with give-and-take on each side. In some cases, mutualistic partners can survive if separated. But in other cases—particularly those in which the partners have become physically adapted to live together—only one partner, or neither, can live without the other.

Mutualism can exist between two animals, between two plants, or between a plant and an animal. Occasionally, there is some overlapping between mutualism and commensalism. When this overlapping occurs, it is not always possible to determine whether only one organism or both benefit from the partnership. But, as the following examples will show, there are many clear-cut instances of mutualism in which both partners plainly benefit from the relationship.

Between Two Animals

zebra and ostrich

Large land animals sometimes join forces for safety. In Africa, for example, zebras and ostriches feed together for protection from the lion and other large cats. The ostrich is the largest flightless bird in the world. Smaller animals do not attack it, for the bird weighs up to 300 pounds and can grow to a height of eight feet. The ostrich has another advantage in its keen eyesight, which enables it to spot large predatory animals long before they get close enough to attack.

The zebra lacks the sharp eyesight of the ostrich. But it has excellent hearing and a keen sense of smell, which the ostrich lacks. When zebras and ostriches graze together in mixed herds, each animal benefits from the other's ability to detect approaching enemies. If either animal senses danger, the entire herd takes off. Since zebras and ostriches both travel at speeds of up to 40 miles an hour, they can outrun any predator except the cheetah, fastest of all land animals.

Ostriches and zebras are among the favorite prey of

the lion. Alone, neither animal has much chance of escaping this fierce hunter. But when the zebra and the ostrich work together for defense, alerting each other to danger, the lion's only chance of catching one of them is by surprise. (Sometimes, though, the cat is able to overpower old, sick, or very young zebras and ostriches when the herd flees.) The zebra and the ostrich thus have an excellent reason—survival—for their mutualistic relationship.

honey guide and ratel

In Africa and Asia, a truly remarkable relationship exists between a bird and a mammal. This is the mutualistic partnership between the honey guide and the ratel.

Brown or gray in color, the honey guide is a small drab bird with white markings on its tail. Usually, the bird feeds on flying insects, especially bees and wasps. But the honey guide also has a strange craving for beeswax. The bird searches out bees' nests with its sharp eyes, often locating them in unused burrows and abandoned termite mounds. The honey guide is a master at finding beehives. But the small bird is not strong enough to tear open the hives to get at the beeswax.

The honey guide is very resourceful, though. After finding a bees' nest, the bird flies off, giving a loud, high-pitched churring call. This call is a signal to let other animals know that the honey guide needs help to

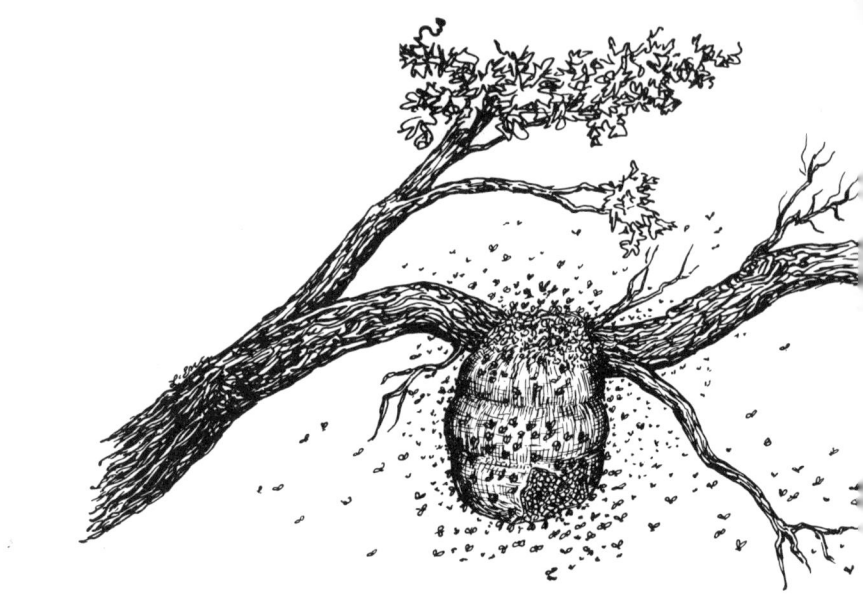

break into the beehive. Sometimes a baboon, or even a human being, answers the honey guide's call. But the partner that usually comes to the honey guide's aid is the ratel.

The ratel is a large badgerlike mammal that has the coloring of a skunk and that protects itself by discharging a foul-smelling liquid similar to that of the skunk. The ratel will eat almost anything—birds, snakes, insects, berries, and roots. In addition, it has a special liking for honey, and will go to almost any lengths to obtain it. This explains why the ratel is sometimes called the "honey badger."

The ratel is not very clever at finding bees' nests. But when it hears the churring call of the honey guide, it takes off in the direction of the call. Once the honey guide sights the ratel, it waits for the mammal to catch up. Then the bird flies off toward the bees' nest, giving its churring call and stopping every now and then to wait for the ratel. In this way, the honey guide leads the ratel to the nest, guiding its partner to honey.

The patient honey guide looks on from above while its badgerlike partner the ratel feasts on honey.

After the honey guide arrives at the bee's nest, it perches on the branch of a nearby tree and waits for the ratel. As soon as the ratel finds the beehive, it digs into it with its long sharp claws and rips the hive open. While the ratel is feasting on the honey, the angry bees try to drive the intruder off. But the bees are not very successful, for the thick, loose fur of the ratel protects it from bee stings.

When the ratel can eat no more honey, it ambles off. Then the patient honey guide swoops down and eats the beeswax that lies scattered about the nest. Through their mutualistic partnership, then, the honey guide and the ratel both get what they want: the honey guide gets its beeswax; and the ratel, its honey. Certainly, theirs is one of the "sweetest" friendships in all of nature!

ants and aphids

In the world of insects, there are many mutual-aid partnerships. One known to all gardeners is that between ants and aphids.

Many kinds of ants like to eat sweet things. If sugar is spilled on the kitchen floor, ants will soon be there. And when you go on a picnic, ants often find the goodies before you do. But ants living in places where no people live have few ways of obtaining the sweets they find so pleasing. One way ants can have a good supply of sweets, however, is by keeping a colony of aphids, or plant lice. The ants raise and care for the aphids much as farmers do their cattle.

Aphids feed by sucking the sweet juice or sap from the tender young shoots of plants. When the aphids take in more sap than they can use, their bodies turn the sap into a sweet, sugary liquid called "honeydew." Ants have learned that aphids excrete this honeydew

An ant milks some aphids for their sweet, sugary honeydew.

from their abdomens when they are stroked on the back with the ants' antennae. So the ants keep herds of the aphids, "milking" them for their honeydew. The ants build stables for the aphids, and protect them from their enemies. So that the aphids can feed on the most nourishing parts of plants, the ants constantly move them from place to place. When winter comes, the ants carry the aphids to underground burrows so that they will not freeze. The aphids' eggs are cared for in the ants' nests during the winter, and the young aphids are carried to good feeding places in the spring.

Both insects benefit from this arrangement. While the ants get their sweet honeydew, the aphids are cared for, sheltered, and protected from their enemies.

termites and protozoa

The close partnership between termites and protozoa (prote-uh-ZOE-uh) is an almost perfect example of mutualism.

Many varieties of termites eat wood, the main substance of which is cellulose. The termites must

39

have wood if they are to survive, but they have no way of breaking down and digesting the cellulose by themselves. Here is where the protozoa come in. These microscopic one-celled animals live in the termites' digestive tracts. They secrete special enzymes that break down the cellulose in wood so that it can be digested by the termites. Like the termites, the protozoa use the cellulose for their food, so that both animals are nourished at the same time.

When they are born, termites do not have any protozoa in their digestive tracts. So they are fed food that already contains protozoa by the adult termites. The protozoa quickly multiply inside the termites' digestive tracts, thus enabling the young termites to digest the cellulose in wood.

Both the termites and the protozoa receive nourishment from their close association, and neither animal can live without the other. Thus they have a good reason for getting together and pooling their resources for the common good.

hermit crab and sea anemone

Many sea animals live together in associations benefiting both partners. One of the best known mutualistic partnerships in the sea is that between the hermit crab and the sea anemone.

Unlike most kinds of crabs, the hermit crab is not born with a hard protective shell. So soon after birth, it starts searching for an empty shell left behind by some other creature such as a sea snail. Once the crab has found a suitable shell, it backs into the shell so that the soft rear half of its body is protected. The hermit crab then moves about freely in search of food, its abdomen snugly fit inside the shell. After the crab has outgrown its borrowed shell, it must move into a larger one. This process is repeated from time to time as the crab continues to grow in size. Thus the hermit crab may live in several different shells—each one larger than the one before it—during its lifetime.

The hermit crab is not really a hermit, for barnacles, sponges, and other sea creatures frequently cling to its shell. One hermit crab of European waters has gone so far as to team up with a sea anemone—a plantlike animal with a fleshy, cylindrical base topped with hundreds of poisonous, petallike tentacles. After the European hermit crab moves into a shell, it immediately starts looking for an anemone partner. When the crab finds a likely looking sea anemone, it grips the anemone with its claws and lifts the anemone up to its shell. In turn, the sea anemone attaches itself to the hermit crab's shell by its fleshy base.

Both animals gain much from this arrangement. The sea anemone cannot move about by itself, but the hermit crab can. When attached to the hermit crab's shell, the anemone gets a free ride and is thus better able to find food. In return, the deadly stinging tentacles of the sea anemone scare away predators that might otherwise eat or attack the hermit crab. When food is caught by either partner, it is usually shared with the other. Thus food, protection, and transportation are the three things that make the underwater partnership of the hermit crab and the sea anemone so binding.

Experiments have shown that neither partner does very well without the other. When the hermit crab outgrows its shell and moves to a larger one, the sea anemone goes, too. It would be impossible to pull the anemone off the crab's old shell if it didn't want to go. But the sea anemone willingly allows the hermit crab to pick it up and transfer it to the new shell. The

The plantlike creature attached to the shell of this hermit crab is a sea anemone.

anemone knows that it has a good thing going for it, and it doesn't want to be separated from its helpful partner, the hermit crab.

giant sea anemone and clown fish

There are many other types of anemones besides the kind that teams up with the hermit crab. One, the giant sea anemone of Australia, is the partner of a small tropical fish known as the clown fish.

Some kinds of sea anemones are only an inch or so in diameter. The giant sea anemone, however, can be anywhere from 16 to 48 inches across. Found in the coral reefs of the South Pacific Ocean, this fascinating creature attaches itself to rocks, shells, and other surfaces. It looks like a beautiful plant as the multi-colored tentacles surrounding its mouth wave back and forth in the water. These graceful tentacles have thousands of poisonous stinging cells that can paralyze and kill other creatures of the sea. Any small fish, mollusk, or crustacean wandering within reach of the tentacles is quickly paralyzed, carried to the anemone's mouth, and eaten. Shells and other parts of the victim that cannot be digested are spewed out of the anemone's mouth opening.

The unlikely partner of the giant sea anemone is the clown fish, a brightly colored tropical fish that measures only three inches in length. Most sea creatures are camouflaged with protective coloring so that they blend into their ocean habitats. Not so the clown

Two clown fish swim among the poisonous tentacles of a giant sea anemone. The brightly colored fish are sometimes called "decoy fish" and "damselfish."

fish. As it swims along, slowly and clumsily, its brilliant stripes of orange, white, and black stand out conspicuously in the dark ocean depths. Predators have no trouble spotting the clown fish. But they learn to avoid it, for they know it associates with the deadly sea anemone.

Unlike other fish, clown fish are not hurt by the stinging tentacles of the giant sea anemone. Some scientists believe that a chemical in the mucus that coats their scales makes the clown fish immune to the sea anemone's poison. Whatever the case, the little fish swim freely and fearlessly among the sea anemone's tentacles, finding shelter and protection there. They venture out to capture food particles and then swim back to safety, to avoid being eaten by larger fish. Besides receiving shelter and protection, the clown fish sometimes snap up scraps of food dropped by the anemone.

The giant sea anemone also benefits from the relationship. For the brightly colored clown fish frequently act as decoys, luring larger fish within reach of the sea anemone's tentacles. Clearly, then, the partnership between the giant sea anemone and the clown fish is one in which *both partners* are helped.

symbiotic cleaning

Symbiotic cleaning is a special kind of mutualistic partnership between two animals. The smaller animal cleans the larger one and helps keep it free from disease. In return, the "cleaner" gets a free meal. Symbiotic cleaning is a widespread type of mutualism that occurs among both land animals and sea animals. Wherever it occurs, this distinct form of mutualism is a truly remarkable phenomenon.

Crocodile and Crocodile Bird. The first symbiotic partnership ever recorded was probably that of the African crocodile and the Egyptian plover, a fearless little bird that is sometimes called the "crocodile bird." This partnership was described almost 2,500 years ago by the Greek historian Herodotus, who wrote about seeing the plover run fearlessly in and out of the crocodile's mouth. Herodotus's description was accurate enough. But not until modern times have scientists been able to understand and explain the strange mutualistic partnership between the crocodile and the crocodile bird.

The warm waters of the African rivers, where the crocodile spends most of its time, are infested with leeches. These small flattened worms attach themselves to the inside of the crocodile's mouth and suck blood. Sometimes the lips, tongue, gums, and throat of the crocodile become covered with leeches. These blood-sucking pests are annoying to the crocodile, and they cause it considerable discomfort.

During the daytime, the leech-ridden crocodiles like to pull themselves out of the water and sunbathe on the mud flats by the river. Often, they lie with their mouths wide open so that the plovers that frequent the mud flats can run in and out. These little gray-and-white birds pick up and eat the leeches covering the inside of the crocodiles' mouths. The crocodile birds also eat small particles of food that are wedged in between the crocodiles' sharp teeth.

Although crocodiles often feed on birds, they are careful not to harm their mutualistic partners, the plovers. The crocodile birds obtain a free meal, and the crocodiles get their mouths cleaned of the bothersome leeches. In addition to eating the leeches and cleaning

The fearless little plover obtains a free meal by cleaning the teeth and mouth of its mutualistic partner, the crocodile.

the crocodiles' teeth, crocodile birds pick off and eat aquatic plants and animals clinging to the crocodiles' thick skin.

Fish-Cleaners and Their Customers. When symbiotic cleaning occurs between two land animals such as a crocodile and a crocodile bird, both partners benefit, but the health of the larger animal is not dependent upon the cleaning services of the smaller animal. In the symbiotic cleaning partnerships of *sea animals*, however, the health—and even the life—of the larger animal is often dependent upon the cleaning services of the smaller animal.

Throughout the oceans of the world are countless cleaning stations where fish come to be cleaned of the harmful parasites that cling to their bodies. These cleaning stations are often located in rocks, coral reefs, kelp beds, and other places that offer protection for the fish-cleaners as they wait for, and work on, their customers. Whereas some fish come to the same cleaning station day after day, other fish seek out a different station each day. Wherever they go, though, the fish have their bodies cleaned of bacteria, fungi, dead and diseased tissue, and various kinds of skin parasites known as fish lice. Many of these parasites are small crustaceans that dig into the skin of the fish, causing sores, and even death, if they are not promptly removed.

There are about 30 kinds of small fish that are known to clean other species of larger fish. Almost all of these symbiotic cleaners are brilliantly colored fish with distinctive patterns of lengthwise stripes that make them visible at some distance. Most fish-cleaners advertise their services to attract customers and to prevent themselves from being mistaken for free meals. Sometimes they rush out and put on elaborate displays to attract new customers. They dart around, swimming back and forth, flipping sideways, nudging their prospective customers, and nibbling at them until the customers stop to be cleaned. Often, the customers go into a sort of trance while being cleaned. Then they float at an odd angle, with their fins extended, so that the fish-cleaners can reach all parts of their bodies. Some cooperative customers lift up their gill covers for the fish-cleaners, and others change their colors so that the parasites, wounds, and patches of bacteria and fungi on their bodies will show up better.

All fish-cleaners have sharp, tweezerlike teeth that aid them in their work. Most fish-cleaners start at the nose of each customer and nibble their way down the body, eating parasites and cleaning wounds and sores. They also clean the gills and gill covers, and often finish the job by entering the mouths of their customers to clean around the teeth and to look for any parasites that may be hiding there. In this way, both the cleaners and their customers benefit: the cleaners get a free meal, and their customers have their bodies cleaned of disease-causing parasites, fungi, and bacteria.

The symbiotic cleaning of fish is the most widespread form of mutualism between animals in the world. Almost all species of marine fish avail themselves of this service, for it is impossible for them not to come in contact with harmful parasites and bacteria. Sometimes, hundreds of fish line up and wait to be cleaned at the same station. The fish-cleaners are smaller than their customers, but larger fish do not try to eat them because of the valuable services they perform. In many ways, the services of a fish-cleaner are as important to the health of its customers as a doctor's services are to the health of his or her patients.

Between Two Plants

There are not many mutual-aid associations among plants. The word *symbiosis*, however, was originally used by scientists to describe the partnership between two unlike plants that grow together to form an entirely different kind of plant. This plant is known as a lichen (LIE-kun), and it consists of an alga (AL-guh) and a fungus (FUNG-gus).

Algae (AL-jee) and fungi (FUN-jy) are both very simple plants lacking roots, stems, leaves, flowers, and seeds. Both kinds of plants are found in all parts of the world, and both can grow where there is no soil. But here, the similarities end. Algae are small green plants that grow mostly in water and that come in the form of seaweed, kelp, and pond scum. These plants contain the green pigment chlorophyll, which enables them to manufacture food through the process of photosynthesis. Some types of algae contain other pigments besides chlorophyll, and come in colors of red and brown. But since *all* algae contain chlorophyll, they are all classified as green plants, regardless of their color.

Fungi are rapidly growing plants that feed on dead and decaying matter or that attach themselves to other living organisms. They come in such various forms as molds, mildews, mushrooms, rusts, and smuts. Unlike algae, fungi have no food-producing chlorophyll. So they frequently team up with algae, thus forming lichens. The fungi obtain their food from their algal partners, but they are not parasites. They help the algae by absorbing and storing water, and by providing the algae with salts, minerals, support—and even protection. For when an alga and a fungus unite to form a lichen, the threadlike strands or fibers of the fungus grow around the alga, forming a tough protective coat.

The lichens formed by algae-fungi combinations are strong, healthy plants that can grow almost anywhere, including inhospitable wastelands where no other plants can grow. Lichens can withstand extreme heat and cold, as well as extreme dampness and dryness;

Lichens, which are made up of algae and fungi, can grow on trees, on tree stumps, and even on rocky, barren ground.

they can live in arctic and tropical regions, on snow-capped mountains and sunbaked deserts. These highly adaptable plants require no soil, and so can grow on rocks, wood, barren ground, trees and tree stumps, and even on fence rails!

Scientists estimate that there are about 16,000 different kinds of lichens. They are of all shapes and sizes, and come in such various colors as green, blue, yellow, brown, gray, and black. All lichens can be roughly divided into three broad groups or classes. Stalked lichens are mossy or shrublike lichens that grow in clusters on the ground or that hang in long strands from the branches of trees. Shell-like lichens resemble shells, and grow in crusty patches on rocks and trees. Leaflike lichens, the third major group, are flat, papery lichens that grow on trees, rocks, and fence rails.

When algae and fungi unite to form lichens, they help not only themselves, but many other living things as well. Some lichens grow on bare rocks and secrete acids that help to break down the rocks into small loose particles. These rock particles, along with decayed lichens, help to form new layers of soil. Other types of lichens growing in bleak arctic regions serve as an important source of food for reindeer, caribou, and other animals, including human beings. Still other types of lichens are used in the manufacture of litmus paper and commercial dyes.

The mutualistic partnership between the alga and

the fungus is one that rivals any such partnership between two animals. Indeed, it is the closest and most successful of all mutual-aid associations. For the alga and the fungus do not merely live together; they grow together and become one, so that what benefits the alga also benefits the fungus, and vice versa.

Between Plants and Animals

Many plants and animals associate in close mutual-aid partnerships. Often, both partners have become adapted in some way for life with the other, and cannot survive without the other's help. If one side of the partnership were to be wiped out, the other side could not exist.

flowering plants and pollen-carrying insects

One of the most important and familiar plant-animal partnerships is that between flowering plants and pollen-carrying insects. When bees and butterflies dive into the blossoms of plants to gather and feed on nectar, they get dusted with pollen. As they fly on from blossom to blossom, they deposit the pollen and so pollinate the plants. At least half of all plant species depend on insects for cross-pollination. Without the help of the nectar-gathering, pollen-carrying insects, most flowering plants could not reproduce and would soon die out.

When we think of insects and cross-pollination, we

think first of the honeybee. This is probably because honey has always been a favorite food of people the world over. For centuries, in fact, honey was the only sweetener most people had. Although there are more than 10,000 different species of bees, only 2 of these produce much honey. But *all* species of bees gather the nectar from flowering plants, pollinating the plants as they move from blossom to blossom.

In trying to get rid of insect pests by spraying deadly chemicals, farmers often kill bees without realizing how important bees are to their crops. More than 100,000 species of flowering plants—including many fruit trees and vegetables—are pollinated by bees. Some orchard owners actually rent hives of bees when their crops are in bloom to ensure pollination. Butterflies and moths are also pollinators. The moth, being a night-flying insect, is especially important in pollinating night-blooming plants.

The widespread partnership between flowering plants and pollen-carrying insects is clearly beneficial to both sides. The insects get the sweet nectar they desire, and the plants are pollinated so that they can reproduce.

birds and plants

There are a great many mutualistic partnerships between birds and plants. Some birds—hummingbirds, sunbirds, and honey creepers—pollinate flowering plants in the same way that bees and butterflies do.

A Gila woodpecker hollows out its nest in a saguaro cactus.

Other species of birds eat the fruit and seeds of trees, shrubs, and grasses. Some of the seeds pass through the birds' digestive tracts and fall to the ground in the birds' droppings. Both sides benefit from this relationship, for the birds are fed, and the seeds are carried to and planted in places away from the parent plants.

Some birds and plants help one another in even more unusual ways. When you see a woodpecker beating out a rat-a-tat on a tree trunk, it may seem rather rough on the tree. The woodpecker, however, is actually helping to rid the tree of harmful insect pests as it secures a meal for itself. So both the bird and the plant benefit from the woodpecker's noisy work.

One particularly interesting woodpecker is the Gila

woodpecker of the American Southwest. This redheaded bird plans ahead. It hollows out its nest in a saguaro cactus a season before it plans to use it. At first, the hole is too moist for a nest. But by the time the woodpecker is ready to use it, the inside of the hole is dry and hard. After the Gila woodpecker moves into the hole, it feeds on the moths that eat and harm the giant saguaro cactus. This way, both the bird and the tree benefit from the partnership. The woodpecker gets a safe nesting place in the spiny cactus, and, in return, helps to rid the cactus of harmful insect pests.

mushroom fungi and leafcutter ants

More than 100 species of leafcutter ants raise their own mushroom-producing fungi in underground gardens. These gardens are from 1 to 15 feet below the ground, and may be as large as 100 square feet.

In parts of Central and South America, and in some southern regions of the United States, these medium-sized red ants are a serious threat to agriculture. In order to obtain material for their gardens, the large worker ants march out each day in search of leaves. When a long file of them goes into a farmer's field to cut leaves, only bare stalks are left behind. Some of the worker ants crawl up the stalks and cut off the leaves. Others stay below, cutting the leaves into smaller pieces that can be easily carried to the nest. Usually, the ants march back to the nest in a long and orderly file, with each ant carrying a small piece of leaf over its head.

Every member of the ant colony does a certain type of work. After the large worker ants have carried the leaves to the nest, they turn the leaves over to smaller workers. These smaller workers spend their whole lives cultivating the underground fungal gardens. They chew up the leaves to make a loose, spongy mixture of saliva and compost for the mushroom fungi to grow in. The growing fungi soon develop hundreds of tiny threadlike strands; and when the ants bite off the ends of these strands, little mushroom bulbs begin to sprout. These mushroom bulbs are the only food eaten by the leafcutter ants.

In this strange mutual-aid partnership, neither side can survive without the other. The mushroom fungi cannot grow without the care and attention of the ants, and the leafcutter ants cannot live without the food produced by the fungi.

three-toed sloth and algae

Another fascinating plant-animal partnership occurs in South America between the three-toed sloth and algae, the tiny green plants discussed in the section on lichens. In this mutualistic arrangement, the sloth provides the algae with a home, and the algae provide the sloth with protective coloration.

The three-toed sloth is a tree-dwelling mammal of South America whose relatives include the armadillo and the aardvark. As its name suggests, it has three long hooklike toes or claws on each of its front feet.

The three-toed sloth is an odd-looking, slow-moving creature whose hooklike claws enable it to hang upside down from the branches of trees.

The sloth uses these claws to hang upside down from the branches of trees. The animal sleeps in this peculiar position during the day. And when night arrives, it walks upside down in search of leaves, buds, and fruits, slowly inching its way along branches with its hooklike claws. Because the sloth is such a sluggish and slow-moving animal, the word *slothful* is sometimes used to describe people who are lazy or slow moving.

The three-toed sloth looks almost as strange as it acts. About 20 inches long, it is a homely creature with a small round head, tiny eyes and ears (the ears are barely visible), a blunt nose, and almost no tail at all. The sloth is covered with coarse, shaggy hair, which is gray or brown in color. The sloth's hair does not always appear gray or brown, however, for algae frequently grow on or in it, making the sloth's hair appear green.

The algae grow very well on the sloth's long thick hair, finding both shelter and protection there. In return, the small green plants help their mutualistic partner by camouflaging it with protective coloration. During the dry season most plants turn brown. Being brown or gray in color, the sloth blends in with its surroundings and is hard to detect. But during the rainy season, when trees and other plants come alive with green, the drab-colored sloth can be spotted quite easily unless algae are growing on its hair. If the algae *are* present, the sloth's hair takes on their green coloration. As a result, the sloth blends in with its green surroundings and so is protected from eagles, jaguars, and other keen-sighted predators. It makes good sense, then, for the sloth to allow its small green friends to get in its hair!

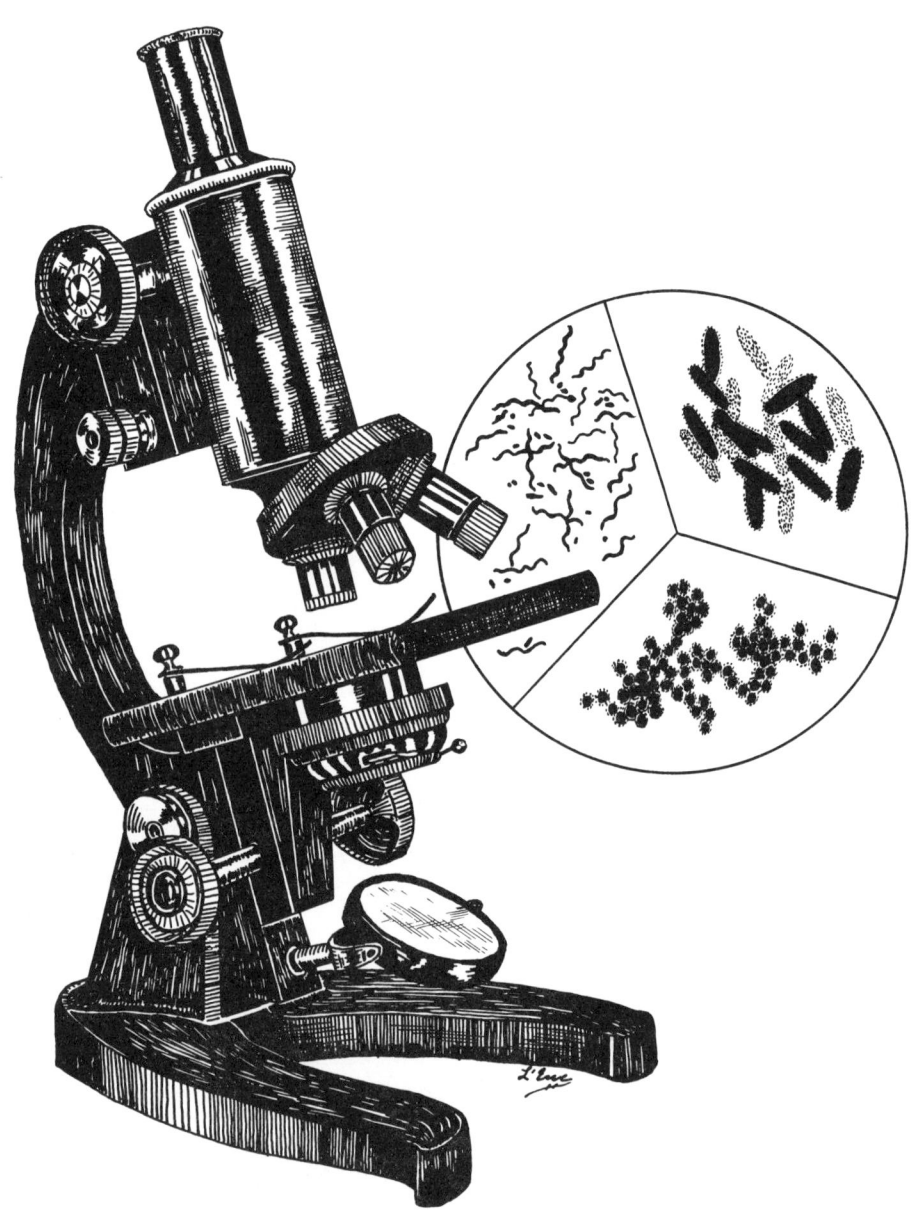

The microscope gives a greatly enlarged view of three different kinds of bacteria, one-celled organisms too small to be seen by the naked eye. Some bacteria are helpful, but others are parasites that can harm and destroy both plants and animals.

PARASITISM

Parasitism, the third major kind of symbiosis, occurs when one organism, the *parasite*, obtains food and shelter from another organism, the host. Nearly all living things harbor some type or types of parasites. The parasite is usually smaller than the host, and it cannot live without the host's help. The host, on the other hand, gains nothing from its one-sided relationship with the parasite, and is often harmed by it.

Some forms of parasitism are more or less permanent, as in the case of tapeworms, which spend most of their lives in the intestines of one animal. Other forms of parasitism are temporary, as in the case of mosquitoes, ticks, and leeches, which feed on their hosts' blood for a while and then leave, never to return. Since most parasites feed only on small amounts of their hosts' tissues and food at a time, they cause little or no harm. An efficient parasite does not intentionally kill its host, for it depends upon the host in order to survive. But some kinds of parasites, such as the hookworms that live in the intestines of human beings, can cause serious damage to their hosts. And other types of parasites—particularly those that bite the skin and feed on blood—can transmit deadly diseases to their hosts. Then, too, whenever many parasites are living off the same host, they may cause the host to become sick or to die.

Parasites come in the form of bacteria, plants, and animals. For their hosts, they may have either plants or animals, or both. Some parasites can live with many different kinds of hosts; and a few spend part of their lives in an *intermediate*, or secondary, host before reaching their *primary* host. But most parasites are physically adapted to live with only one or a few different hosts. Whatever the case, however, the nature of the relationship is the same: the parasite always takes, and the host always gives—involuntarily, and sometimes at the cost of its life.

Parasitic Bacteria

As you learned on page 23, bacteria are neither plants nor animals, but they can have the qualities of both. These microscopic one-celled organisms are among the smallest of all living things: a million bacteria live in one drop of saliva, and a billion in one gram of dirt. Most kinds of bacteria are beneficial, and there would be little or no life on earth without them.

But some kinds of bacteria act as parasites, causing serious diseases in plants and animals. These parasitic bacteria cannot easily move about by themselves. But they are often carried to their hosts by insects and other small animals.

plant diseases caused by parasitic bacteria

About 200 different types of parasitic bacteria produce diseases in plants. These bacteria cause three different kinds of damage to plants. They can enter the stems of plants and prevent the movement of water, causing the plants to wilt. They can kill plant cells and so cause spots on fruits, vegetables, and flowers. And they can produce a lumpy overgrowth, or gall, on plants. Often, two or more types of parasitic bacteria attack a plant at the same time. When this happens, the plant has little chance of surviving.

Fire blight was the first plant disease known to be caused by parasitic bacteria. It is a serious disease of pears in areas of North America where the summers are warm and rainy. Fire blight has destroyed all the high-quality pear orchards in some parts of the country, causing the pear trees to wither and die. The disease also affects apple and quince trees, along with various other plants.

Fire blight bacteria are spread by wind, rain, and bees. As bees travel from flower to flower, they carry the bacteria along with the pollen. The bacteria enter the blossoms and quickly kill the flower cells. Then,

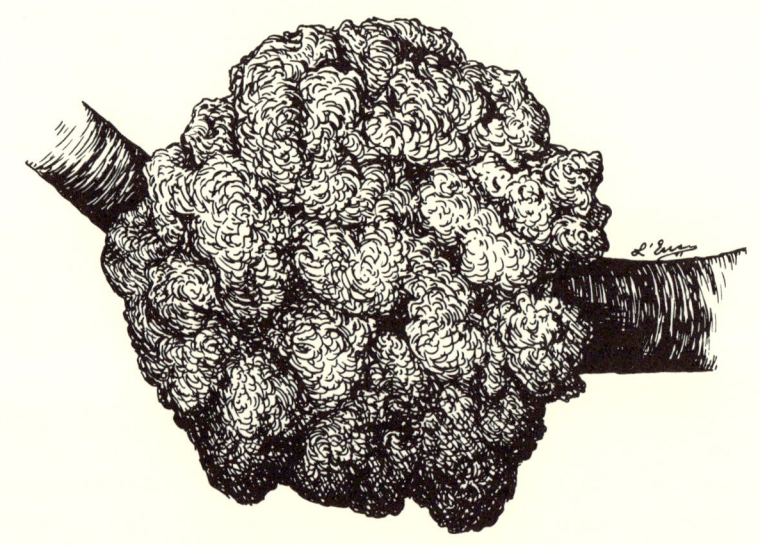

Some parasitic bacteria cause plant cells to increase rapidly in size and number, producing the lumpy growth called a gall.

after spreading to the young fruits, stems, and leaves, the bacteria travel down to the roots and kill the tree. Diseased areas of trees hit by fire blight produce an oozy substance that is loaded with bacteria. This substance helps spread the disease to other trees.

Crown gall is perhaps the best known bacterial plant disease. The parasitic bacteria that cause this disease can only enter a plant through a wound. They are carried from plant to plant by chewing insects, and they are washed into the soil from diseased plants. When invaded by the bacteria, plant cells rapidly increase in size and number, forming the knobby growths known as galls. In all, crown gall affects more than

40 plant families, including roses, sugar beets, peaches, apples, raspberries, grapes, figs, and pecans.

animal diseases caused by parasitic bacteria

Most parasitic bacteria of human beings and other animals get their nourishment by digesting the blood, muscles, and other tissues of their hosts. Diseases are caused by the damage done to the tissues and by the poisonous wastes released into the bloodstream by the parasites.

Parasitic bacteria are spread to animals in several ways. Some are airborne, and can be taken into the respiratory tract with air. Others enter the alimentary canal with food and water. Many parasitic bacteria are spread through insect bites, and a few types grow on the dead tissues in wounds.

Certain airborne bacteria attack the moist linings of the nose, mouth, and throat, causing diseases there. These bacteria are spread by coughing, sneezing, or spitting. Diphtheria, whooping cough, scarlet fever, and tonsillitis are spread in this way. So are the airborne bacteria that cause tuberculosis and pneumonia, two of the most serious lung diseases of human beings.

Parasitic bacteria are often taken into the alimentary canal with food and water. Bacteria can be left on an object by someone who has a disease and then be picked up by another person. Sometimes food is contaminated with parasitic bacteria carried by common houseflies. Typhoid fever and bacterial dysentery can both be spread this way. Tuberculosis of the bones, glands, and organs other than the lungs can be contracted by drinking milk that contains tuberculosis bacteria.

The terrible plague caused by the bacterium *Pasteurella pestis* was the great killer of the Middle Ages. The plague is primarily a disease of rats. But the bacteria that cause the disease can be transmitted to human beings by the fleas from infected rats. Usually, the plague does not affect people. Whenever a large number of rats have been killed in an epidemic, however, the fleas have quickly spread to human beings, infecting them with the disease-causing bacteria. During the Middle Ages several plague epidemics swept through Europe and other parts of the world. (In the 1300s a severe form of the plague called the "black death" wiped out a fourth of the population of Europe.) But today, thanks to strict quarantines, modern sanitation measures, and vigilant rat-extermination programs, the plague has been nearly stamped out.

Parasitic Plants

Some plants, unable to manufacture their own food, have turned to a life of parasitism. There are two major classes of parasitic plants—the fungi, and the seed plants. Parasitic fungi, which prey on either plants or animals, are small; parasitic seed plants, which prey only on other plants, can be quite large.

parasitic fungi

Fungi, you will recall, are very simple plants lacking stems, roots, and leaves. They also lack chlorophyll, and so are unable to manufacture their own food. In all, over 100,000 species of fungi have been discovered and classified. Some of these fungi live and feed on dead or decaying matter. But many types of fungi are parasites that get their nourishment from other living organisms—both plants and animals.

Fungal Diseases of Plants. All fungi reproduce by forming tiny spores, which are scattered about by the wind. Since one fungus produces thousands of spores, fungal colonies spread very rapidly. After the spores of a parasitic fungus fall on the leaves or stems of a plant, they germinate, or start to grow and develop. In many cases, a tiny tube soon sprouts from the spores. This tube searches about until it finds an opening by which it can enter the host plant. Then the fungus forms a hairlike network inside the plant as it sucks nourishment from it. Some parasitic fungi have tiny needlelike points with which they can pierce the tissues of plants. Once they have found their way inside another plant, the fungi can do their hosts much harm, robbing them of food and minerals, and ravaging them with diseases.

About 20,000 different kinds of parasitic fungi cause diseases in plants. These fungal diseases include smuts, rusts, mildews, and blights. Nearly 75 percent of all infectious plant diseases are caused by fungal blights alone, which cost farmers millions of dollars in crop losses each year.

In 1930 a parasitic fungus began attacking Dutch elm trees, widely used as shade trees in the eastern and central United States. Dutch elm disease causes the tissues of elm trees to become clogged, preventing food and water from reaching the leaves. The parasitic fungus that produces this disease is usually carried from tree to tree by the elm bark beetle, which breeds only in Dutch elm trees. Since 1930, Dutch elm disease has spread across the country, destroying millions of magnificent elm trees.

As you can see, a tiny parasitic organism like a fungus can affect an entire country. Sometimes the effects of a fungal disease are felt by millions of people. During the 1800s the poor peasants of Ireland were almost totally dependent on their potato crops for food. Beginning in 1845, the potato crops in Ireland were affected by the late blight fungus. This disease spreads rapidly in the cool, moist regions where potatoes are grown. In Ireland the fungus caused both stored potatoes and those growing in the fields to rot. The spores of the fungus survived the winter, and the fungal blight broke out again when potatoes were planted in the spring.

For three straight years—1845, 1846, and 1847—the potato crops in Ireland failed. About 1 million people (one out of every eight) died of starvation. Another 1.5 million (one out of every five) emigrated to the United States to escape starvation. Many years later, during World War I, late blight fungus hit the potato crops in Germany. Many people believe that this fungal disease helped to shorten World War I by reducing essential supplies of food.

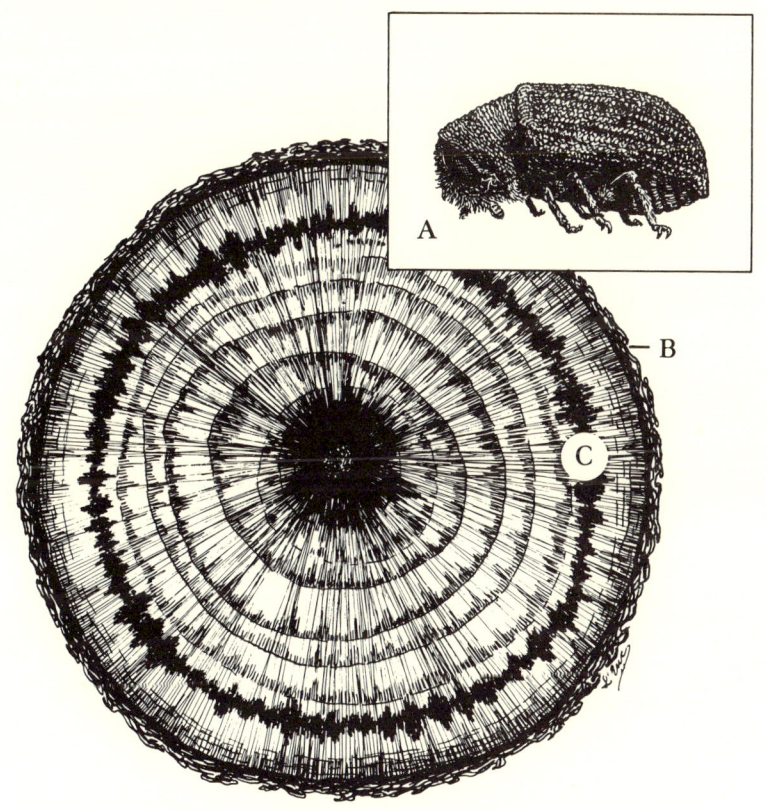

The elm bark beetle (A) carries the parasitic fungus that causes Dutch elm disease. The beetle lays its eggs in the bark of the tree (B), and the beetle's larvae tunnel their way to the tree's water-conducting vessels. Once here, the disease-causing fungus grows until it blocks off the passages (C).

Fungal Diseases of Animals. Parasitic fungi can attack animals as well as plants. In fact, all animals, including human beings, are host to some types of fungi. Parasitic fungi that attack animals usually enter the host through an opening in the body such as a cut, a scratch, a wound, or an insect bite. Once they have entered the host, the fungi can do considerable harm, causing a wide variety of diseases.

People are especially concerned with fungal diseases that attack domesticated animals. Histoplasmosis, which affects the lungs, spleen, and central nervous system, attacks both wild and domesticated animals. San Joaquin Valley fever, caused by a fungus that lives in the respiratory tract, produces a serious cough in cattle, sheep, and dogs. Lumpy jaw, another fungal disease, attacks pigs, cattle, and horses.

Some of the fungal diseases of domesticated animals, such as San Joaquin Valley fever, can also attack people. There are few serious fungal diseases of human beings, however, most of them coming in the form of mild skin diseases.

Ringworm is a fungal skin disease that children often get by playing with infected cats. The fungi that cause ringworm enter the skin around the hair follicles. They destroy the hair, which then falls out, leaving ring-shaped patches on the scalp. These patches itch, and the fungi are spread through scratching. The many spores produced by the fungi help to spread the disease over the skin of the host. When the host comes in contact with other people, the fungal spores spread to them, infecting the new hosts with ringworm.

Another fungal disease of the skin is athlete's foot. The parasitic fungi that cause this disease thrive in warm, damp conditions, and are usually found on the floors of baths, swimming pools, locker rooms, and gymnasiums. People pick up the fungi by walking barefoot on infected floors. The parasitic fungi grow best between the toes, where they cause intense itching and burning. As you have probably guessed, this highly contagious disease is called "athlete's foot" because it is most often picked up by athletes from infected locker room floors.

parasitic seed plants

Fourteen species of flowering, seed-producing plants are parasites of other plants. Some depend on their hosts for all of their needs. Others are semiparasitic, furnishing themselves with at least some of the things they need.

Mistletoe is one of the best known of the semiparasitic seed plants. The leaves of the mistletoe contain chlorophyll; so the plant can carry out photosynthesis to manufacture its own food. But since the mistletoe lacks true roots, it has no way of getting the water and minerals it needs. Here is where the host plant comes in. Whenever the mistletoe plant comes in contact with a tree, tiny rootlike organs start to grow. These rootlike organs grow into the branches of the host tree, enabling

Mistletoe is a parasitic seed plant that can be identified by its small green leaves, its yellowish flowers, and its shiny white berries.

the mistletoe to suck up the water and minerals it requires.

Small yellowish flowers appear on mistletoe plants in February and March. These are followed by shiny white berries. Birds love to eat the berries, which often stick to their beaks. When eaten, the berries pass through the birds' digestive tracts unharmed, and then stick to the birds' feathers. The birds land on trees to try to wipe the sticky berries off their feathers and beaks. The outside pulp of the berries usually sticks to the trees and hardens, protecting the seeds until they germinate.

Mistletoe plants, although slow growing, are persistent and long lived. These parasites do not usually die unless their hosts die. In the Temperate Zone, mistletoe plants do little damage to their hosts. But in the warm Tropical Zone, where mistletoe plants require a great deal of water, they sometimes kill their hosts by draining them of all their water and minerals.

Unlike the mistletoe plant, the dodder is a parasitic seed plant that lacks chlorophyll and so is entirely dependent on its host for food, minerals, and water. After the seeds of the dodder plant germinate in the ground, a thin stemlike tendril grows up. This tendril grows toward the nearest green plant and soon encircles the plant's stem. Within a few days, rootlike structures with sucking organs grow from the dodder plant's tendril into the host plant. The real roots of the dodder then drop off, leaving the parasite completely dependent on its host for food and water. Often, the host plant is so preyed upon by the "devil's sewing thread," as the dodder is sometimes called, that the host dies.

Another parasitic seed plant is the broomrape. This parasite attacks the wild broom, as well as alfalfa and other cultivated plants. It attaches itself to the roots of the host plant with special sucking organs, through which it robs the host of water, minerals, and food. Great damage is done to fields of clover and alfalfa when they are invaded by the broomrape, which is one of the most destructive of all the parasitic seed plants.

Parasitic Animals

At one time, many parasitic animals probably lived independently, without the need of hosts. But over the centuries, they adapted to a parasitic way of life in which food could be obtained directly from the tissues, waste products, and partially digested food of other organisms. As these animals adapted to a life of parasitism, their bodies gradually changed. Many of them lost their wings, legs, and other means of locomotion as they lost their need to move about. Their alimentary canals became simpler, or disappeared altogether. At the same time, organs used by the parasites to attach themselves to their hosts and to suck and absorb nourishment from them became more and more specialized.

There are three broad classes of parasitic animals: protozoa, or one-celled animals; worms; and insects, ticks, and mites. Some parasitic animals, including protozoa and tapeworms, live inside their hosts and are known as *endoparasites.* Others, like leeches and aphids, live on the skin or outside surface of their hosts and are known as *ectoparasites.*

Most parasitic animals prey only on other animals. Some prey on plants, however, and a few can parasitize both plants and animals. Nevertheless, all parasitic animals are limited as to the number of different hosts they can prey on. Some are adapted to only one host, whereas others can use any of several similar hosts. Then, too, some parasitic animals pass one or more cycles of their life in an intermediate host before reaching their primary host.

parasitic protozoa

As you learned in the preceding chapter, protozoa are one-celled animals that can be seen only under a microscope. There are more than 20,000 different kinds of protozoa, and almost all of them live in water. Many kinds of protozoa are helpful, for they serve as an important source of food for fish and other sea animals. But some kinds of protozoa act as harmful endoparasites, causing serious diseases in human beings and other animals. Often, these parasitic protozoa are carried by an intermediate host to a primary host.

Amebas (ah-ME-bahz) are perhaps the simplest of all protozoa. Most kinds of amebas are harmless; but one, *Entamoeba histolytica,* is a parasite that lives in the intestinal tract of human beings, causing a serious disease known as amebic dysentery. Amebic dysentery is found all over the world, but it is most prevalent and serious in tropical countries. Scientists estimate that at any given time, close to 10 percent of the world's population has amebic dysentery.

During their dormant, or resting, stages, the parasitic amebas that cause amebic dysentery form protective saclike structures known as cysts. The amebas are enclosed in these cysts when they pass out in the feces of infected individuals. Other people become infected with the amebas when they drink water or eat uncooked food containing the cysts. After the cysts are swallowed and carried into the intestines, the amebas break out of the cysts and tunnel into the lining of the intestines, feeding on the intestinal tissues of the host. Ulcers, or holes, appear where the amebas have tunneled, and bloody diarrhea usually results. Sometimes the parasitic amebas are carried to the liver or brain, where they may cause further damage and even death.

Several species of parasitic protozoa cause malaria in human beings, monkeys, birds, and reptiles. Different species of protozoa attack different species of mosquitoes, which in turn pass the parasites on to other hosts. The only known species of mosquito that carries and spreads the malaria protozoa to human beings is the *Anopheles.* The female *Anopheles* mosquito needs blood to produce fertile eggs; so she feeds on the blood of human beings. The mosquito acts as the primary host of the protozoa; human beings and other animals serve as the intermediate hosts.

When a female *Anopheles* mosquito sucks blood from a person infected with malaria, the malaria protozoa contained in the person's red blood cells travel into the mosquito's stomach. There, the parasites develop and multiply rapidly. Thousands of the parasitic protozoa eventually make their way into the mosquito's salivary glands. When the mosquito bites another person, it injects saliva containing the parasites into the new host. The parasitic protozoa then attack the person's red blood cells, infecting the host with malaria.

Left: Anopheles *mosquito*

Right: *tsetse fly*

Malaria causes violent attacks of chills, fever, and sweats, along with extreme weakness. As the parasitic protozoa grow and multiply in the host's red blood cells, the cells rupture, releasing the parasites into the bloodstream. This rupturing of the red blood cells and releasing of the protozoa into the bloodstream causes the repeated, regular attacks of chills, fever, and sweats. The loss of red blood cells resulting from the destructive action of the parasites causes weakness and anemia—a serious shortage of red blood cells.

Malaria is common all over the world, and there have been more cases of it than of any other infectious disease. An age-old disease, malaria was first described back in the fifth century B.C. by the Greek physician Hippocrates. Yet not until 1898 did scientists discover that the female *Anopheles* mosquito transmits malaria to human beings. Since then, the eradication of the mosquitoes has been one of the most widespread and effective means of controlling the disease. Quinine and other drugs are used to treat malaria, and antimalarial drugs that destroy the parasitic protozoa can be taken to prevent catching it.

Another deadly disease caused by parasitic protozoa is trypanosomiasis, commonly known as "sleeping sickness." This disease has been spread by the tsetse fly, a bloodsucking gnat, over 4 million square miles of tropical Africa. The parasitic protozoa that cause the disease attack the host's brain, causing headaches, chills, and fever. As the parasites destroy more and more of the central nervous system, the victim goes into a deep and prolonged sleep that often ends in death. Sleeping sickness attacks domesticated animals as well as human beings, making it almost impossible for farmers to raise cattle. Fortunately, the disease does not spread outside of Africa because the tsetse fly occurs only there.

parasitic worms

Worms make up one of the largest classes of parasitic animals. Most parasitic worms live inside their hosts, and so are classified as endoparasites. But a few live as ectoparasites, attaching themselves to the outer surface of their hosts and sucking blood from them. The following section will discuss three of the most important groups of parasitic worms: the roundworms, the flatworms, and the leeches.

Roundworms. There are thousands of different species of the slender, cylinder-shaped worms known as roundworms, and a great many of them live as parasites. Nearly 100 different kinds of eelworms—tiny, threadlike roundworms—attack plants, feeding on their tissues and destroying their leaves, stems, and roots. But an even larger number of parasitic roundworms attack human beings and other animals, causing serious diseases among them.

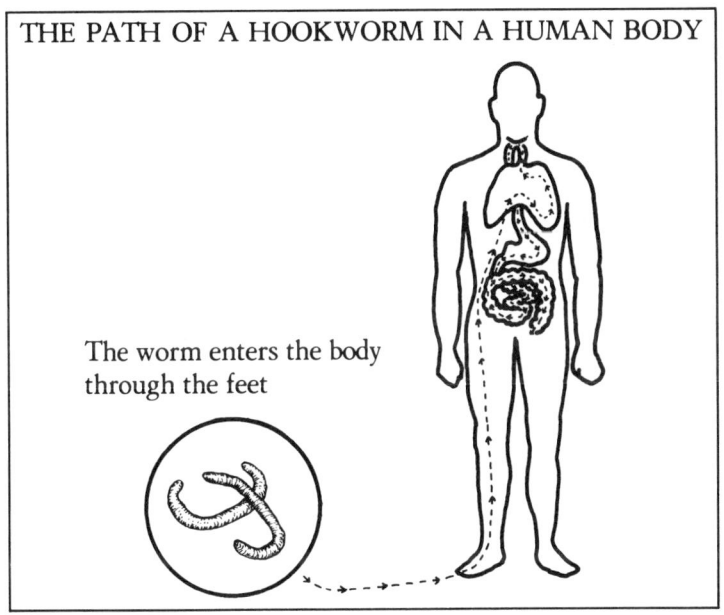

THE PATH OF A HOOKWORM IN A HUMAN BODY

The worm enters the body through the feet

Hookworms are the most harmful of the parasitic roundworms that attack human beings. The tiny larvae of these worms hatch in warm, moist soil. They enter their hosts by burrowing through the skin, and are most often picked up by people who walk barefoot in tropical countries. Once they are in the host's body, the hookworm larvae enter the bloodstream and migrate to the lungs. From the lungs, they travel up the respiratory tract to the throat, where they are swallowed into the intestines. Using little hooks on their mouths, the larvae then attach themselves to the intestinal walls and feed on the blood and body fluids of the host. After the larvae have developed into adult hookworms, they lay thousands of eggs, which are passed out in the host's feces. If left untreated, hookworm disease can cause weakness, abdominal swelling, diarrhea, and chronic anemia.

Hookworms occur only in tropical and subtropical countries, where the soil is warm and moist. A parasitic roundworm of cooler climates is the pinworm, which is only about one-fourth of an inch long. Although this small white worm lives in the intestines of human beings, it lays its eggs on their skin. Pinworm infection spreads rapidly among children, who often swallow the pinworm eggs when they eat without first washing their hands. Pinworm infection can be very annoying, for the eggs can cause intense itching and swelling of the skin. The disease is rarely dangerous, however, and it can be treated with a number of drugs.

Another parasitic roundworm is the trichina (trih-KY-nuh) worm. This endoparasite lives in the muscles of hogs, rats, and human beings, causing a painful disease known as trichinosis (trick-uh-NO-siss). People get the disease by eating undercooked pork containing the larvae of the small trichina worm, which is less than a fourth of an inch long. After the larvae enter a person's body, they break out of their saclike cysts and travel to the intestines, where they grow into adults. The female trichina worms then burrow into the intestinal wall and produce large numbers of new larvae. The larvae quickly enter the host's bloodstream and travel throughout the body. Eventually, they form cysts in the muscles of the neck and chest, causing muscular

Trichina worm in cyst lodged in muscle

pain and making breathing difficult. Other symptoms of trichinosis include fever, headache, nausea, swelling of the face and neck, and bleeding under the skin. Fortunately, the parasitic roundworms that cause this painful disease can be killed by freezing infected pork before it is cooked or by cooking the meat thoroughly before it is eaten.

Flatworms. A flatworm is a slender worm with a flattened body. Many flatworms are endoparasites that live inside the bodies of human beings and other animals. The two major groups of parasitic flatworms are the tapeworms and the flukes, both of which can do considerable harm to their hosts.

Tapeworms are long ribbonlike flatworms that live in and feed on the intestines of human beings and other animals, including fish, hogs, and cattle. The body of a tapeworm is composed of a chain of thin, blocklike segments. Tapeworms grow from the neck back: each new segment formed at the neck pushes the others back, so that the segments farthest from the neck are the oldest. New segments are formed all the time, and some tapeworms reach a length of 30 feet or more.

Tapeworms attach themselves to their hosts' intestines with little suckers and hooks on their heads. The hungry parasites have no mouths; but they absorb food from the intestines through their body walls. Some adult tapeworms lay millions of eggs every day. Eventually, these eggs pass out of the host's digestive tract with the feces. When a fish or a pig or a cow eats food

Tapeworms can reach a length of 30 feet or more.

containing the tapeworm eggs, the eggs live as larvae in the muscles and other organs of the infected animal. If a person later eats undercooked meat containing the tapeworm larvae, the larvae will travel to the new host's intestines, where they will grow into adult tapeworms.

The human hosts of tapeworms are usually thin and hungry, no matter how much they eat. In addition to weight loss and irregular appetite, tapeworms cause such ill effects as weakness, anemia, abdominal pain, and nervousness. But tapeworms almost never kill their hosts, and they can be expelled from their hosts' intestines with several different drugs.

Unlike tapeworms, which have many body segments, flukes are parasitic flatworms having only *one* body segment. There are many kinds of flukes, each attacking different hosts and each invading different parts of the body. In the Temperate Zone, flukes often infect cattle, sheep, and dogs, damaging their livers. In the Far East and in some tropical regions of the Western Hemisphere, flukes commonly infect human beings, living in their intestines, blood, lungs, and livers.

Many flukes live in intermediate hosts during one stage of their development. For example, after the eggs of blood flukes hatch in the water, the larvae often live in water snails. Then, after the larvae have developed and multiplied, they leave the snails and enter the bodies of fish, crabs, and other animals. Sometimes the larvae of blood flukes burrow into the skins of swimmers and travel to the veins and blood vessels. Here, they feed on their hosts' blood and develop into adult flukes.

Leeches. Leeches, or bloodsuckers, are one of the few groups of worms that live as ectoparasites. Using large disklike suckers, they attach themselves to the skin or outer surface of their host. Then they penetrate the host's skin with their teeth and suck out the blood. Since most leeches live in water, their most common hosts are fish, frogs, and turtles. But, as you learned in the last chapter, leeches will also attack crocodiles. And when given the chance, they will attach themselves to human beings. While they are sucking blood from their hosts, leeches secrete a special chemical substance that prevents the blood from coagulating, or thickening. This makes it easier for the leeches to obtain the nourishment they must have in order to survive.

Leech

parasitic insects, ticks, and mites

Parasitic insects, ticks, and mites make up the last major class of animal parasites. Although some of these parasites attack plants, most of them attack other animals, feeding on their blood and infecting them with diseases.

Some Prey upon Plants. The aphid, or plant louse, is probably the best known of all the parasitic insects that prey upon plants. This soft-bodied insect is so small—about one-sixteenth of an inch long—that it often goes unnoticed. Yet a colony of the tiny parasites can quickly infect an entire garden or orchard, damaging or destroying all the plants in its path. Using their tube-shaped mouths, aphids pierce the stems of plants and then feed on the plants' juices and sap. When sucking the nourishment from their hosts, the parasites inject into the plants a poisonous substance that discolors and sometimes deforms them. Later, after the aphids have gone, parasitic bacteria and fungi sometimes enter the puncture wounds made by the aphids and proceed to do further damage to the plants.

Aphids multiply rapidly, but their numbers are kept under control by their natural predators. Spiders, ladybugs, and beetles destroy large colonies of aphids in both their larval and adult stages. Aphid lions, or lacewings, also prey upon aphids, keeping their populations in check.

Some species of parasitic wasps, flies, and gnats cause the growth of harmful galls on plants. These insects pierce the leaves, stems, and roots of certain host plants and then deposit their eggs in the tissues of the plants. When the eggs hatch, the plant tissues surrounding the larvae begin to swell rapidly, forming lumplike galls. The larvae inside the galls feed on the hosts' juices until they have grown into adults. Then they burrow small holes in the galls and fly away. One gall-making insect, the Hessian fly, destroys millions of dollars worth of wheat crops every year.

Scale insects make up the largest group of parasitic insects that feed upon plants. The bodies of these small round insects are covered with scaly shells, which explains their name. Scale insects have no legs or eyes or feelers. But they have tubelike beaks through which they can suck out the juices from fruit trees, shrubs, and other plants. There are more than 2,000 different kinds of scale insects, and all of them attack plants in large numbers. One scale insect, the cottony cushion scale, came close to destroying all the citrus trees in California. This harmful parasite was finally brought under control in 1890 when its natural predator, the ladybird beetle of Australia, was imported into California.

Some Prey upon Other Animals. Thousands of different species of insects, ticks, and mites prey upon human beings and other animals. Many of them are ectoparasites that bite the skin of their hosts, drawing up blood and body fluids. Others are usually classified as endoparasites, for they burrow into their hosts' skin.

Many ectoparasitic insects have mouthparts that are especially adapted to pierce the skin of other animals. Mosquitoes, for example, have six needlelike mouthparts that they use to puncture the skin and to sip the blood of their hosts. The bites of mosquitoes, gnats, and other bloodsucking insects are often painful, and they can cause itching and swelling. But far more serious are the diseases that these parasites can transmit to their hosts. As you learned earlier, the *Anopheles* mosquito carries malaria; and the tsetse fly, a gnat, transmits sleeping sickness. Three other diseases transmitted by the bites of mosquitoes are yellow fever (a tropical disease caused by a virus), encephalitis (a viral disease that attacks the brain), and elephantiasis (a skin disease characterized by the enlargement of the legs).

Flea

Fleas are small wingless insects that live on cats, dogs, birds, rabbits, horses, rats, and human beings. Like mosquitoes, these bloodsucking pests are well suited to a life of parasitism. Their strong, bristly legs enable them to move easily through the feathers and hairs of their hosts. They have sharp beaks to pierce the skin, and tubelike mouthparts for sucking blood. Since fleas store blood in their abdomens, which are quite large, they can survive without food for weeks while searching for new hosts. Their quests for new hosts seldom take that long, however, for most fleas can attack any number of different animals.

Flea bites cause intense itching. Some hosts develop an allergic reaction to the bites, and the skin around the bites becomes inflamed. A severe infestation of fleas, especially on a small host, can cause weakness from loss of blood, and even death. But much more serious is the threat the fleas pose to human beings. For fleas, like mosquitoes, are important carriers of diseases. When they bite human hosts, they can infect them with disease-causing bacteria, protozoa, and viruses. The fleas from infected rats, you will recall, transmit the plague to human beings. Other fleas act as the principal carriers of typhus (an infectious disease characterized by fever, headache, and red spots on the skin) and tularemia (a bacterial disease characterized by fever, aching, and inflammation of the lymph glands).

The louse is another small wingless insect that feeds on the blood of other animals. This crablike parasite has strong claws, which it uses to cling to its hosts, and beaklike mouthparts adapted for piercing and sucking. Some lice, known as biting lice, chew on the feathers and skin of birds. These lice will also attack raccoons and horses. Other types of lice—particularly head, body, and crab lice—attack human beings, causing serious skin and scalp irritations. Body lice are the most harmful of these parasites, for they spread epidemic typhus (a serious and widespread kind of typhus) to human beings.

Biting lice on feather

Ticks and mites are small oval-shaped creatures belonging to the same scientific class as spiders and scorpions. They are not insects, for they have eight legs instead of six. But like mosquitoes, fleas, and other bloodsucking insects, they are tiny parasites that prey upon a large number of animals, including human beings. Ticks and mites are considered endoparasites, for they burrow into their hosts' skin, where they gorge themselves on blood and body fluids. About the only important difference between them is their size: ticks are larger than mites, and they can usually be seen with the naked eye.

Female ticks lay thousands of eggs at a time. The eggs hatch on the ground, and the larvae crawl up blades of grass to wait for hosts. When suitable animals pass by, the larvae descend on them. Then they burrow into the hosts' skin with their toothed beaks, feeding on the hosts' blood and body fluids.

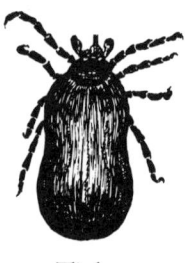

Tick

Among the animals most frequently attacked by ticks are cattle, sheep, chickens, dogs, and human beings. Ticks can harm their hosts by releasing poisonous substances into the bloodstream. If a host is small enough, the poisonous bite of a tick can cause paralysis. And if a tick is forcibly pulled out of a host, its head may break off and remain inside the skin, causing a bad infection. More serious, however, are the diseases that are transmitted by tick bites. These include relapsing fever (a bacterial disease of human beings), Texas fever (a protozoa-caused disease of cattle), and Rocky Mountain spotted fever (a disease of human beings that is caused by bacterialike microorganisms).

Mite

Mites are smaller than ticks, but they are no less harmful or annoying. They attack the same hosts as ticks, and transmit the same diseases. Some of the mites that attack human beings—particularly chiggers, or "red bugs"—burrow deep into the skin, causing the skin to itch and to break out in rashes. Other mites burrow into the hair follicles and oil glands of human beings. But no matter where they enter their hosts, and no matter which hosts they enter, all mites are after the same thing: blood. For without the lifeblood of other animals, these tiny parasites cannot live.

※ ※ ※ ※

We are apt to think of mites, ticks, and other parasites as being "bad," because they live off other living things without helping them, and because they often harm and sometimes even destroy their hosts. But in nature, there are no good or bad forms of life. Each living thing has adapted to life in a certain niche where it can secure food and shelter in order to survive and to reproduce its kind. Since parasites are among the lower forms of life, they have probably been forced to team up with higher, more self-sufficient forms of life. In adapting themselves to their hosts, and in finding ways to obtain nourishment and shelter from them, they have responded to the most basic instinct of all living things. And that is the instinct of self-preservation.

Glossary

algae. Simple green plants containing chlorophyll and other pigments (especially brown and red), and having no true roots, stems, or leaves. Found in all parts of the world, algae grow mostly in water and damp places, and come in the form of kelp, seaweed, and pond scum. When algae unite with fungi, lichens are formed.

bacteria. One-celled organisms so small that they can be seen only with a microscope. Although bacteria are neither plants nor animals, they can have the qualities of both. Some bacteria are helpful, such as the ones that enrich the soil by breaking down dead and decaying matter. But others are harmful parasites that attack both plants and animals, causing serious diseases among them.

commensalism. A kind of symbiosis in which one organism, the "guest," is helped by the relationship while the other organism, the "host," is neither helped nor harmed by it. In this loose, one-sided arrangement, the guest receives food, shelter, transportation, or protection from the host, which is usually the larger of the two. Commensal partners are not dependent upon each other, and, if separated, both survive.

ectoparasites. Parasites that live on the skin or outer surface of their hosts. Aphids, leeches, mosquitoes, and gnats are all classified as ectoparasites.

endoparasites. Parasites that live and feed inside their hosts, inhabiting their internal organs and tissues. Among the endoparasites that can invade the bodies of human beings are hookworms, pinworms, tapeworms, and flukes.

fungi. Simple plants lacking chlorophyll and having no true roots, stems, or leaves. Some fungi unite with algae to form lichens, and some enrich the soil by breaking down dead and decaying matter. But other types of fungi are parasites that attack plants and animals, robbing them of food and ravaging them with diseases.

lichen. The small, highly adaptable plant that is formed when an alga and a fungus team up and live together as one. The alga produces food for the fungus, which in turn provides the alga with water, minerals, and protection.

mutualism. A kind of symbiosis in which two organisms team up for mutual aid, forming a close, two-sided partnership. Both of the partners benefit from the relationship, with give-and-take on each side. Mutualism can exist between two animals, between two plants, or between a plant and an animal.

parasitism. A kind of symbiosis in which one organism, the "parasite," is helped by the relationship while the other organism, the "host," is harmed by it—or even destroyed by it. Parasites are usually smaller than their hosts, and they cannot live without them. Whereas some parasites spend their whole lives with a single host, others spend part of their lives with an "intermediate," or secondary, host before reaching their "primary" host.

phoresy. A type of commensalism in which one organism obtains transportation by clinging to a larger organism of a different species. The smaller of the two is called the "phoront," and the larger the "host."

protozoa. One-celled animals too small to be seen with the naked eye. Most kinds of protozoa live in water. While many of them are helpful, some protozoa act as disease-causing endoparasites that attack human beings and other animals.

symbiosis. The living together of two unlike organisms. Sometimes the partnership is between two animals, sometimes between two plants, and sometimes between a plant and an animal.

symbiotic cleaning. A mutualistic form of symbiosis in which one animal cleans another and so helps keep it free from disease. In return for its services, the "cleaner" obtains a free meal.

synoecy. A type of commensalism in which one organism—the guest—receives shelter from another one—the host— which is usually the larger of the two.

Index

algae: definition of, 50; and fungi, 50-53; and three-toed sloth, 57-59
amebas, 75-76
amebic dysentery, 75-76
anemia, definition of, 77
Anopheles mosquito, 76-77, 86
ants, 7; and aphids, 38-39; army, 16-17; black tree, 28-29; leaf-cutter, 56-57
aphid lions, 84
aphids, 74, 84; and ants, 38-39
army ants, 16-17
athlete's foot, 71

bacteria, 48, 49, 60; airborne, 65; definition of, 23, 62; and flies, 23; parasitic, 60, 62-66, 84, 87, 88
bacterial dysentery, 65-66
barnacles, 23-24, 41
"beaters," 13, 14, 16
bee-eater, 13-15
bees, 7, 14, 35-37, 53, 54, 63
beetle: dung, 21-22; elm bark, 68, 69; ladybird, 85
birds, 7, 8, 72 (*see also under individual names*); and army ants, 16-17; and grazing animals, 10, 12; and plants, 54-56
biting lice, 87
"black death," 66
black tree ants, 28-29

blights: bacterial, 63-64; fungal, 67-69
blood flukes, 83
bloodsuckers, 46, 83
body lice, 87
broomrape, 73

carmine bee-eater, 13-15
cattle egret, 10, 12
cellulose, 39-40
chiggers, 88
chlorophyll, 50, 67, 71, 73
clingfish, 31
clown fish, 43-45
commensalism, definition of, 8, 11
cottony cushion scale, 85
crab: hermit, 40-42; pea, 30, 31
crocodile, 45-47, 83
"crocodile bird," 45-47
cross-pollination, 53-54
crown gall, 64-65
cysts: of parasitic amebas, 75-76; of trichina larvae, 80-81

diphtheria, 65
diseases, 48, 49, 62, 75-89 (*see also under individual names*); bacterial, of animals, 23, 63, 65-66; bacterial, of plants, 63-65; carriers of, 63-64, 65-66, 68, 69, 75, 76, 77-78, 83, 85-87, 88; fungal, of animals, 70-71; fungal, of plants, 67-69

93

dodder, 73
dung beetle, 21-22
Dutch elm disease, 68, 69
dysentery: amebic, 75-76; bacterial, 65-66

Echeneis naucrates, 20
ectoparasites, 74, 78, 83, 85
eelworms, 78
egret, 10, 12
elephantiasis, 86
elm bark beetle, 68, 69
encephalitis, 86
endoparasites, 74, 75, 78, 80, 81, 85, 87-88
Entamoeba histolytica, 75
epidemic typhus, 87

fire blight, 63-64
fish, 7, 20, 25, 31, 32, 43, 45 (*see also under individual names*); symbiotic cleaning among, 47-49
fish-cleaners, 47, 48-49
fish lice, 48
flatworms, 81-83
fleas, 66, 86-87
flies: and bacteria, 23, 65-66; Hessian, 85; parasitic, 84, 85; tsetse, 77-78, 86
flowering plants, 53-54
flukes, 81, 82-83
fungi, 48, 49; and algae, 50-53; definition of, 50; mushroom, and leafcutter ants, 56-57; parasitic, 67-71, 84

gall, 63, 84-85; crown, 64-65

giant sea anemone, 43-45
Gila woodpecker, 55-56
gnats, 77, 84, 85-86
goby, 30, 31
ground hornbills, 16-17
guenon monkey, 15
guest, commensal, 11

hat-pin urchin, 31-32
hermit crab, 40-42
Hessian fly, 85
histoplasmosis, 70
"honey badger," 35, 36-38
"honeydew," 38-39
honey guide, 35-38
hookworms, 62, 79-80
hornbills: ground, 16-17; pompadoured, 15
host: commensal, 11; *intermediate*, 62, 75, 76, 82-83; in a parasitic relationship, 61-62; *primary*, 62, 75, 76

innkeeper worm, 30, 31
insects, 7, 10, 12-14, 16, 28, 35, 55-56, 63, 65 (*see also under individual names*); parasitic, 74, 83-87; pollen-carrying, 53-54
intermediate host, 62, 75, 76, 82-83
Irish potato famine, 68-69

kori bustard, 13-14

ladybird beetle, 85
larvae: mite, and dung beetle, 21-22; of parasitic insects,

84-85; of parasitic worms, 79, 80-81, 82, 83
late blight fungus, 68-69
leafcutter ants, 56-57
leeches, 46, 61, 74, 83
lice. *See under individual names*
lichen, 50-53; definition of, 50
louse, 87
lumpy jaw, 70

malaria, 76-77, 86
mistletoe, 71-73
mites, 74, 87-89; larvae of, 21
monkeys, 7; guenon, 15
"monkeybird," 15
mosquitoes, 61, 76, 85-86; *Anopheles*, 76-77, 86
moth, 54
mushroom fungi, 56-57
mutualism, definition of, 8, 33

osprey, 25
ostrich, 34-35

paradise parrot, 26
parasite, definition of, 61-62
parasitism, definition of, 8, 61
Pasteurella pestis, 66
pea crab, 30, 31
phoresy, 19-24; definition of, 19
phoront, definition of, 19
photosynthesis, 50, 71
pilot fish, 17-18, 19
pinworm, 80
plague, the, 66, 87
plant lice, 38, 84
plants: and birds, 54-56; flowering, and pollen-carrying insects, 53-54; seed, parasitic, 66, 71-73
plover, 45-47
pneumonia, 65
pollen-carrying insects, 53-54
pompadoured hornbill, 15
primary host, 62, 75, 76
protective coloration, 14, 43, 57, 59
protozoa: definition of, 40, 75; parasitic, 74, 75-78; and termites, 39-40

ratel, 35, 36-38
rats, 66, 87
relapsing fever, 88
remora, 18, 19-20
rhinoceros, 10, 12
rhynchocephalians, 27
ringworm, 70-71
Rocky Mountain spotted fever, 88
roundworms, 78-81
rufous woodpecker, 28-29

saguaro cactus, 55, 56
San Joaquin Valley fever, 70
scale insects, 85
scale worm, 30, 31
"scarab," 21-22
scarlet fever, 65
sea anemone, 40, 41-43; giant, 43-45
sea urchin, 31-32
seed plants, parasitic, 66, 71-73
shark: and pilot fish, 17-18, 19; and remora, 18, 19-20
shrimpfish, 31

"sleeping sickness," 77-78, 86
sloth, 57-59
sooty shearwater, 27, 28
spores, fungal, 67, 68, 70-71
symbiosis, definition of, 8
symbiotic cleaning. 45-49
synoecy, 24-32; definition of, 24

tapeworms, 61, 74, 81-82
termites: and parrots, 26; and protozoa, 39-40
Texas fever, 88
three-toed sloth, 57-59
ticks, 61, 74, 87-89
tonsillitis, 65
trichina worm, 80-81
trichinosis, 80-81
trypanosomiasis, 77

tsetse fly, 77-78, 86
tuatara, 27-28
tuberculosis, 65, 66
tularemia, 87
typhoid fever, 65-66
typhus, 87; epidemic, 87

"urchin fish," 31-32
Urechis caupo, 31

woodpecker: Gila, 55-56; rufous, 28-29
worms: and dung beetle, 22; parasitic, 74, 78-82. *See also under individual names*

yellow fever, 86

zebras, 7, 34-35